AF545217

Agricultural Transformation and Country Perspectives

Agricultural Transformation and Country Perspectives

Edited by

Subir Ghosh

2012

Icfai Books
The Icfai University Press

AGRICULTURAL TRANSFORMATION: CONCEPTS AND COUNTRY PERSPECTIVES

Editor: Subir Ghosh

First Edition: 2012
Printed in India

Published by

This book is published by IUP.
University Campus, Agartala-Simna Road,
P.O. Kamalghat Sadar, Agartala – 799210, Tripura (West)
E-mail: info@iupindia.org
Website: www.books.iupindia.org

ISBN: 978-81-314-2704-0

Contents

Overview

Agricultural transformation is the process by which individual farms shift from highly diversified, subsistence-oriented production to more specialized production, oriented towards the market or other systems of exchange. The process involves a greater dependence on input and output delivery systems and increased integration of agriculture with other sectors of the domestic and international economies. Agricultural transformation is a broader process in which an increasing proportion of economic output and employment are generated by sectors other than agriculture. For instance, in early 1990s, after the first democratic elections, Romania embarked on a process of general economic reform, which also included specific and ambitious reforms in the agriculture and food sector. The reform program envisaged the transformation of agriculture to a sector based on the principles of private ownership of land and other agricultural property. The aim was to create a market-oriented and internationally competitive agriculture. Since the collapse of Communist system in 1989, Bulgarian agriculture has gone through an unprecedented transformation from a centrally planed to a market based private economy. The fundamental transition has affected significantly agricultural specialization and farming structures (Bachev and Tsuji, 2001; Bencheva, 2005; OECD,

2000). The development challenge that Africa faces is arguably at a critical watershed point. There is increasing global commitment to assist Africa in addressing its problems of poverty, debt and integration into the global marketplace. At the same time, there is increasing controversy over how to do it. Most analysts agree that agricultural transformation is central to redressing Africa's poverty problem. Since 1990, India too realized to shift its agricultural production system to a more specialized market-oriented one.

Against this backdrop, this book focuses to make agricultural production system more specialized and thereby create market-oriented and internationally competitive production system.

This book is structured into the following two sections:

Section I: Introduction, which introduces the concept of agricultural transformation through achieving more specialized and market-oriented and internationally competitive agriculture.

Section II: Country Perspectives, which highlights the experiences of implementing agricultural transformation in different country perspectives.

The first article, "**Agricultural Transformation: Concepts and Prospects**", by *Subir Ghosh* gives an overview of agricultural transformation. It discusses the concept and need of agricultural transformation and also highlights its acceptability in cross country experiences. Agricultural transformation is a process through which a single farm shifts from traditional production system to a highly specialized production system towards market-orientation. Therefore, it needs structural transformation of agricultural sector through the adoption of improved techniques of production. As agricultural transformation ensures to increase agricultural productivity and at the same time helps to produce qualitative and competitive production, it makes agricultural sector more revenue generating source. In other words, it brings attention of rural youth to get involved more in this sector. More involvement of rural youth in agricultural activities helps on the one hand, to solve unemployment problem and on the other hand, solves the food demand of a nation. To make successful achievement of agricultural transformation process, rural youth participation is required as agricultural transformation needs to transform traditional production system to more productive and market-oriented production system. But today's younger generation are not keen to involve in agricultural activities as it is considered a dirty and financially less rewarding work. Therefore, young people are generally migrating to nearby cities or towns for seeking jobs and due to heavy population pressure, job markets get saturated before the rural young people reach them. Therefore, they remain unemployed and are vulnerable to anti-social elements. On the other hand, the legal/genuine question arises if rural people are not attracted to agricultural operations then who will cultivate the agricultural land and how will the food problem of the country be solved? The only way is to make the agricultural production system more productive and market-oriented so that the farmers can sell their product at internationally competitive market in high prices which means successful achievement of agricultural transformation. To make agricultural sector more productive and competitive, proper training facilities with modern devices like computer, Internet, etc; market information, wide market opportunities, agricultural loan, crop insurance, etc should be provided to the farmers. These opportunities may attract the rural people to involve in agricultural operations as it may ensure guarantee to earn good income from agricultural sector. To provide such facilities to the farmers, the concern government's role can not be overemphasized. Therefore, the government has to be actively involved in providing the required facilities for agricultural operation and at the same time, government should encourage nationalized banks for providing loans, private bank like ICICI Lombard for providing agricultural insurance, the private bodies like ITC (Indian Tobacco Company) and NGOs (Non-Government Organizations), etc .,to train the farmers for successful achievement of agricultural transformation.

The second article, **"Farmer Participation in Private Sector Agricultural Extension"**, by *Dominic Glover* examines how far agricultural development has shaped the behavior of the private sector. The private sector has attained a new prominence in policy debates about agricultural development in recent years. In search of affordable and effective ways to inject new energy and dynamism into agricultural extension services, various countries have explored policy options including privatization, greater involvement of private sector service providers, and various models of cost-recovery or fee-charging. To some extent, private commercial provision of agricultural advice and services implies a degree of responsiveness and accountability on the part of the suppliers towards the customers Private sector organizations are increasingly important players in the provision of agricultural extension/transformation services, as well as crop research, and extension that would put the "farmer first" (Chambers *et al.,* 1989). Since then, farmer participatory approaches have been refined, developed, evaluated and critiqued, and have influenced the practice and rhetoric of research and extension. For this purpose, this article uses the case of the Monsanto Smallholder Programme (SHP) to focus specifically on the issues of farmer participation, responsiveness and accountability, in a situation where extension services were being supplied by a transnational company that is also a major producer and marketer of herbicides as well as the dominant driver behind the development and commercialization of genetically modified crops internationally. The paper also explores how the Smallholder Programme was conceived, designed and implemented, discusses how far it was designed around and responded to farmers' needs and priorities, and considers the extent to which farmers were able to hold the company to account.

The third article, "**Effects of Agricultural Management on Soil Organic Matter and Carbon Transformation – A Review**", by *Xiaobing Liu, S J Herbert, A M Hashemi, X Zhang* and *G Ding* summarizes how agricultural transformation, crop rotation, residue and tillage management, fertilization and monoculture affect soil quality, Soil Organic Matter (SOM) and carbon transformation. Soil is a vital natural resource that is non-renewable on the human time scale and is a living, dynamic, natural body that plays many key roles in terrestrial ecosystems. It is the essence of life and health for the well-being of humankind and animals and the major source of most of the food production. The maintenance of soil health is essential for sustained productivity of food, the decomposition of wastes, storage of heat, sequestration of carbon, and the exchange of gases. Soil Organic Matter (SOM) is the central indicator of soil quality and health, which is strongly affected by agricultural management. Soil Organic Matter (SOM) is a major terrestrial pool for carbon, nitrogen, phosphorus and sulpher, and the cycling and availability of these elements are constantly being changed by microbial immobilization and mineralization. The importance of increased SOM or Soil Organic Carbon (SOC) is its effect on improving soil physical properties, conserving water, and increasing available nutrients. These improvements should ultimately lead to greater biomass and crop yield. The results confirm that Soil Organic Matter (SOM) is not only a source of carbon but also a sink for carbon sequestration. Soil is a vital natural resource that is non-renewable on the human time scale (Jenny 1980) and is a living, dynamic, natural body that plays many key roles in terrestrial ecosystems. It is the essence of life and health for the well-being of humankind and animals and the major source of most of our food production. The maintenance of soil health is essential for sustained productivity of food especially through transformation of agriculture, the decomposition of wastes, storage of heat, sequestration of carbon, and the exchange of gases. However, only a limited area of the soil can actually be used for growing food, and when improperly managed it can be eroded, polluted or even destroyed (Brady and Weil 2000). Soil Organic Carbon (SOC) is the most often reported attribute and is chosen as the most important indicator of soil quality and sustainable agricultural transformation.

The fourth article, **"Agricultural Transformation: Towards Marketization"**, by *Nirbachita Karmakar* focuses on different aspects of agricultural transformation like structure of the agrarian systems, describing the process of agricultural transformation and specialization, marketing of the agricultural produce and marketing information system. It describes the shift from subsistence farming to a diversified or mixed farming and finally going for agricultural specialization in the entire process of agriculture transformation. The study brings out the initiatives that can be taken to address the agricultural transformation towards marketization in the context of developing countries. It also highlights that agricultural transformation is acutely prevalent in industrial nations and evolving at par with structural transformation of the national economy. Improvement in the standard of living, biological and technological innovations, and expansion of local, national and international markets for a strong impetus on agricultural transformation. In agricultural transformation, when going for crop specialization, securing food for the family with some marketable surplus no longer remains the basic objective. Rather, the basic goal rests purely on commercial profit with maximum per-hectare production; i.e., production is entirely market oriented. Agricultural transformation emphasizes on capital formation, technological progress, scientific research and economic development. Agricultural transformation varies in terms of size and function with regard to specialized farms. Specialization ranges from cultivated fruit, vegetable farms to the vast stretch of wheat and corn fields with usage of modern laborsaving machinery like tractors, combined harvesters to airborne spraying techniques. Since agricultural transformation is based on specialization, most of the specialized farms lay emphasis on monocrop cultivation, usage of capital intensive technology, and relying on economies of scale to cut back costs of production and focus on profit maximization. Agricultural transformation in a way is more similar to operations carried out in large industrial enterprises. Finally, it talks about one of the most important aspects of marketization, i.e., marketing information system with an illustration in the context of Indian agriculture.

The fifth article, **"Ecosystem Approaches to Research on Agricultural Transformation, Nutrition and Human Health"**, sourced form *www.idrc.ca* a document of *International Development Research Centre, Ottawa*, Canada emphasises that the aim of ecosystem approaches on agricultural transformation and human health is to deepen the understanding of the relationships between the two with a specific emphasis on food security, dietary diversity and their implications on nutrition. Agricultural transformation means a range of changes in food production practices, such as the intensified use of inputs or the introduction of new practices and technologies. Agricultural practices, in particular, play an important role in creating supportive conditions for human health as many aspects of farming practices influence both rural and urban population health. Currently, in many developing countries, agro-ecosystem management practices are in a process of transformation due to increasing population pressure, decreasing availability of agricultural lands, environmental constraints (such as climate change), and contamination from industrial and agricultural activities, increasing urban market demands and globalization processes. Agricultural transformations, especially when coupled with economic and environmental degradation, can adversely impact the relationships between people and the ecosystems on which they depend for their livelihood, affecting patterns of human health, disease and nutritional status. The approaches recognize that there are inextricable links between humans and their biophysical, social, and economic environments that are reflected in individual/communities' health. It focuses on understanding (i) the interactions between social and ecological systems in defining key determinants of human health in particular settings, and (ii) the impact of human activities on the sustainability of these processes. It also seeks to identify ecosystem management strategies that contribute to improving the health and living conditions of human populations and the sustainability of the ecosystem in which they live. Eco-health represents a process-oriented and dynamic way of understanding and solving problems, which can be constructed in various contexts, with varying scales, and different intended outcomes. "Ecosystems" in this approach are defined relative to the research problem and refer to the social and ecological contexts, both on a temporal and a spatial scale, of human lives. Human activities (or stressors) alter these contexts and have positive and negative effects on individuals and communities involved.

The sixth article, **"Agricultural Transformation and Changing Status of Women"**, by *P Narasimha Rao* and *R Venkata Rao* examines the status of women across the levels of rural transformation led by agricultural development. Women play socially and economically productive roles but their social and productive roles are governed by values, norms and customs of the society, which is seen to have placed them in the lower social and economic status. The role and status of women are altered in the process of agrarian social transformation due to the technology of cultivation. Agriculture has been the main source of livelihood for millions of people living in rural areas. Agricultural development brings about sustainable livelihood to the rural poor and would improve the socio-economic and political status and norms. Women play an important role in the agricultural sector. Their role in production, processing, storage and marketing is well established. Agricultural growth consequent upon the globalization of economy altered the organization of production, patterns on ownership of land and cropping patterns, which in turn altered the structure of inequalities in rural economy. This change in the technology and organization of production altered the multiple roles of women and consequently the status of women in the family and society. In this context, it is appropriate to study the economic, social and political roles and changing position of rural women consequent upon the technical changes and agricultural development in the process of globalization of Indian economy. Due to the advancement of agricultural transformation, the position of women, both in the family and in society has radically changed. Therefore, the rural transformation consequent to agricultural growth enabled the women to have access to education and productive resources. The agricultural growth has not only led to a decline in infant mortality rate and maternal mortality rate, but also led to changes in the pattern of gender gap and literacy among women. Further, this study concludes that agricultural transformation brought changes in the health and marriage practices of women as well as political participation among women.

The seventh article, **"Post-Communist Transformation in Bulgaria – Implications for Development of Agricultural Specialization and Farming Structures"**, by *Hrabrin Bachev* highlights a new inter-disciplinary methodology of the New Institutional and Transaction Costs Economics. Since, the collapse of Communist system in 1989 Bulgarian agriculture has gone through an unprecedented transformation from a centrally planed to a market based private economy. The fundamental transition has affected significantly agricultural specialization and farming structures. Therefore, the aim of this paper is to fill the gap and examine modes and factors of post-communist agricultural specialization and farming structures development in Bulgaria. Firstly, it presents the specific Bulgarian model for farming transformation characterizing with restitution of farmland in real borders and original locations, physical distribution of assets of ancient public farms into individual shares, rapid liberalization of markets and prices, and lack of public support to agriculture. Secondly, it specifies factors for evolution of new farm structures and specialization such as badly specified and enforced property rights; big institutional, market and behavioral uncertainty; high assets specificity and dependency; lack of managerial experience; low incentives for long-term investment; ineffective public interventions, etc. Third, it demonstrates how these factors affect organization and specialization of farming in the country explaining the evolution of a huge subsistence and part-time farming, production cooperation at a large scale, unprecedented concentration of resources in few business farms, widespread use of informal and integrated modes etc. Fourth, it analyzes the impact of transition on farm structures and agricultural specialization through changes in structure and share of agricultural GDP (Gross Domestic Product) and employment, and distribution of activities between different types of farms. Finally, it clarifies efficiency of and extends of specialization in dominating large business farms, production cooperatives, and numerous small-scale unregistered farms.

The eighth article, **"Agricultural Extension in Africa and Asia"** by *Carl K Eicher* attempts to review the role of agricultural extension systems in helping smallholder farmers in Africa and Asia increase agriculture production and their livelihoods. The paper also tries to explain the reasons why agricultural extension is an important framework to incorporate into the information and knowledge systems for small farmers. Extension reforms are carried out in the agriculture sector in many countries of Asia and Africa. Finally, it summarizes that the Farmer Field School model is an important institutional innovation that needs to be studied in-depth in different agro-ecological zones, different institutional arrangements and over time. Because of the lack of baseline data and adequate monitoring of ongoing FFS (Farmer Field School) activities at the farmer and community levels, the available evidence suggests that it is premature to promote the FFS model as the "best model" for developing countries. It will take time and resources for researchers to study and evaluate this important institutional innovation. Meanwhile, international organizations such as the FAO should endorse the concept of the pluralism of extension models, including the FFS models. Countries should be encouraged to collect data on the impact, costs and returns of the FFS model, including its financial sustainability, in a learning-by-doing manner.

The ninth article, "**Agricultural Transformation in Latin America**" by *Basistha Chatterjee* reviews the progress of agricultural transformation in Latin America. Agricultural transformation is an innovative as well as integrated approach which rightly aims at (i) to increase in productivity and efficiency at all stages of the commodity chains, for greater value addition and employment generation and (ii) to reduce the cost of transaction between different stages of agribusiness system. Although transformed agriculture is the overall picture of Latin America but the degree of transformation is varied within the ten countries namely Brazil, Argentina, Bolivia, Chile, Colombia, Ecuador, Paraguay, Peru, Uruguay and Venezuela. Poverty and inequality is still persisting at a high degree. In spite of 30 per cent increase in agriculture value added the rural poverty rate is unchanged in last ten years. Farmers are still feeling that the subsistence agriculture is the supporting cushion for them. The wave of global market demand for value added and newly derived agricultural produce has touched Latin America to a great extent. Latin America has also the big supply potential especially the countries like Brazil, Chilie, Peru and Guatemala. This paper therefore envisages the challenges on agricultural transformation in Latin America which is led by global market force. But the transformation of agriculture is not homogeneous in Latin America. Only the skillful farmers and the modernized processing sector are enjoying the fruits of transformation while the others are still stuck with traditional farming. Indeed, new agricultural inputs, new agro-technology have facilitated this transformation but unless the small farmers can enter into the arena it cannot be said to be complete and healthy because small farmers encompass the big scenario in the agricultural sector of Latin America. It is also a challenge because of the fact that, in spite of agricultural transformation the level of poverty and unemployment remained static during the last ten years. The paper also suggests the possible solutions of the problem.

The tenth article, **"Agricultural Transformation: The Case of China"** by *Asis Kumar Pain* discusses the progress of agricultural transformation in China. The People's Republic of China under the leadership of Mao Zedong, is poised to emerge as the largest economy in the world in about two decades (by the year 2015), overtaking both Japan and the USA. The article looks into the economic and social development of new China by reaping the full advantages of surplus labour in rural areas that leads to sharp rise in the rate of capital formation in the economy as a whole without any undue restriction of the rate of rise of mass consumption. The article thereafter highlights the contribution made by land reform in China towards utilizing rural resources for financing industrial development. Also the innovativeness instituted in the agricultural production is given a look. In rural areas, capital formation is expressed, in terms of physical indicators like the volume of earth shined in reclamation and terracing work, the number of irrigation reservoirs constructed, the miles of canals dredged, etc. On the basis of the discussion, the article opined that that the fundamental innovation of the Maoist development strategy, of agricultural transformation lay in converting an apparent liability into an asset: by directly transforming under-employed agricultural implements into its optimum state at minimal extra cost.

The eleventh article, **"Agricultural Transformation in India"** by *Basistha Chatterjee* and *Visvarup Chakravarti* expresses that with the advent of the Green Revolution in India, agricultural transformation was rapidly visible in 1960-70 mainly boosted by high yielding variety seed of wheat and rice. Afterwards the GDP (Gross Domestic Product) growth continued steadily especially at 4 per cent during 1992 to 1996 and 2 per cent during 1997-2003. Share of rural poor and people below poverty level has also reduced which is a noticeable feature. The subsidizing policy on fertilizers and electricity for irrigation also facilitated the rapid transformation after 1970 but on the other hand pressure on the government budget also exerted simultaneously. To increase the yields the public sector is providing fertilizers, power and water for irrigation at subsidized rate to the farmers. Also a lot of public investments in agricultural research, extension and infrastructural development were done. The national agricultural policy formulated by the Government of India fixes minimum support price for the major commodities and update of prices is done each year for the same. The State Governments are empowered to provide irrigation, power and fertilizers in the respective states. In this way, the public stock disbursement among low income consumers at subsidized prices is maintained through the Public Distribution System (PDS). On food subsidy, India spends a huge portion of her budget. These efforts of increased production as well as the price policies generated a steady growth in agriculture since the Green Revolution in 1960s. By the restrictive policies imposed by the Government the domestic market was insulated from the wave of global trade. Therefore, the paper emphasizes on the consequences of this transformation viz., improved health, education, infrastructure and land use at an early stage. This normally varied from the plains (of the state Punjab, Andhra Pradesh, Karnataka, Maharashtra) to the hilly tracts of Himalayas. The paper also talks about the upcoming challenges – further prosperity, need for employment, limited natural resources etc.

The twelfth and last article, **"Contract Farming and Horticulture: A Perspective on Agricultural Transformation in India"** by *Rahul Gupta* focuses how horticulture helps in agricultural transformation providing an ideal backdrop for practice of contract farming in India. Though India is a leading producer of several food items, a majority of Indian farmers, especially the small and marginal ones, continue to be economically backward. The horticulture sector, where small holding farms are common, is ideally suited for contract farming which would be beneficial to the farmers through agricultural transformation. Contract farming binds the farmer and the sponsor in a business deal with a win-win situation for both the parties, where the projects are not motivated mainly by the political or social considerations, like in the Green Revolution, but is tied up with economic and technical realities. In the area of food grains, export potential of the famous basmati rice has drawn several exporters towards practice of contract farming in order to ensure quality and quantity required by the overseas market, which shows the potential of agricultural transformation in India. Contract farming has had successful applications in consolidating many small farmers and getting the raw material required for the processing plants. In the Indian context, horticulture provides an ideal backdrop for practice of contract farming for selected crops.

Section I

Introduction

1

Agricultural Transformation
Concepts and Prospects

Subir Ghosh

Agricultural transformation is a process through which a single farm shifts from traditional production system to a highly specialized production system towards market orientation. Therefore, it needs structural transformation of agricultural sector through the adoption of improved techniques of production. For this purpose, proper training facilities with modern devices like computer, Internet etc; market information, wide market opportunities, agricultural loan, crop insurance, etc., should be provided to the farmers. In this context, concerned government has to be aware to look into this matter.

Introduction

Agricultural transformation is a process through which a single farm shifts from traditional production system to a highly specialized production system towards market orientation. It heavily depends upon input and output delivery system. Agricultural transformation not only creates output for final consumption but italso supplies inputs for other manufacturing industries. Therefore, agricultural transformation aims to create market-oriented and internationally competitive agricultural production system through the adoption of new technologies so as to improve the socioeconomic status of the rural masses.

In other words, structural transformation also refers to agricultural transformation and defined as the method through which individual farms go for a specialized production from a diversified subsistence production with the objective of creating a more market-oriented agrarian economy. An agriculture and employment based strategy should encompass at least three basic components – a high rate of agricultural production through technological, institutional and price incentive mechanisms and by raising the productivity of small farmers; following an employment based urban development strategy can raise the domestic demand for agricultural output; lastly the farming community can be supported through a diversified non-agricultural and labor-intensive rural activities. Thus agricultural transformation can be market-oriented production system through the structural transformation of rural masses and it will achieve an integrated rural development.

Specialization and Marketization of Agricultural Product: Prospects

Agricultural transformation refers to the process of specializing and making the agricultural produce towards market-oriented. Now the question arises as to which method or technique will be adopted to make the agriculture sector more productive, and at the same time qualitatively productive, because qualitative production only can ensure guarantee to face international competition and help to sell the agricultural produce at higher prices. It also needs wide market opportunities to sell agricultural produce. To get maximum yield from agricultural sector, the involvement of agri-business houses are also required as they can support the marginal farmers through the provision of qualitative seeds, fertilizers, etc., and also help the farmers to dispose agricultural produce at best market prices. Therefore, corporate contract farming is one of the best options for achieving agricultural transformation. In this context, eSAGU system also has a significant contribution towards delivering market information, latest technological advancement, etc., to the farmers through agricultural experts. Figure 1 gives an outline of eSAGU system and how it helps the farmer to make agricultural sector more productive and market-oriented. "ESAGU is an ICT based System where the latest technological advancements are harnessed and adapted to deliver the agro-advisories to the farmers. The ICT is deployed as a tool to ensure timeliness, cost-effectiveness and acts as an enabler to provide other allied services like Agri-Finance, Marketing, Input Supplies and Insurance to farmers on the eSAGU Platform"[1].

Besides eSAGU system, ITC's [Imperial (now renamed as Indian) Tobacco Company] 'Choupal Sagar' has also significant contribution in marketing the rural products. ITC has built new infrastructure through supplementing the presence of e-Choupal – rural marketing hubs called 'Choupal Sagar'. ITC's setting up of rural mall 'Choupal Sagar' at Sehore in Madhya Pradesh is a burning example in this context (Ghosh Subir, 2008).

Figure 1: The Parts of eSagu System

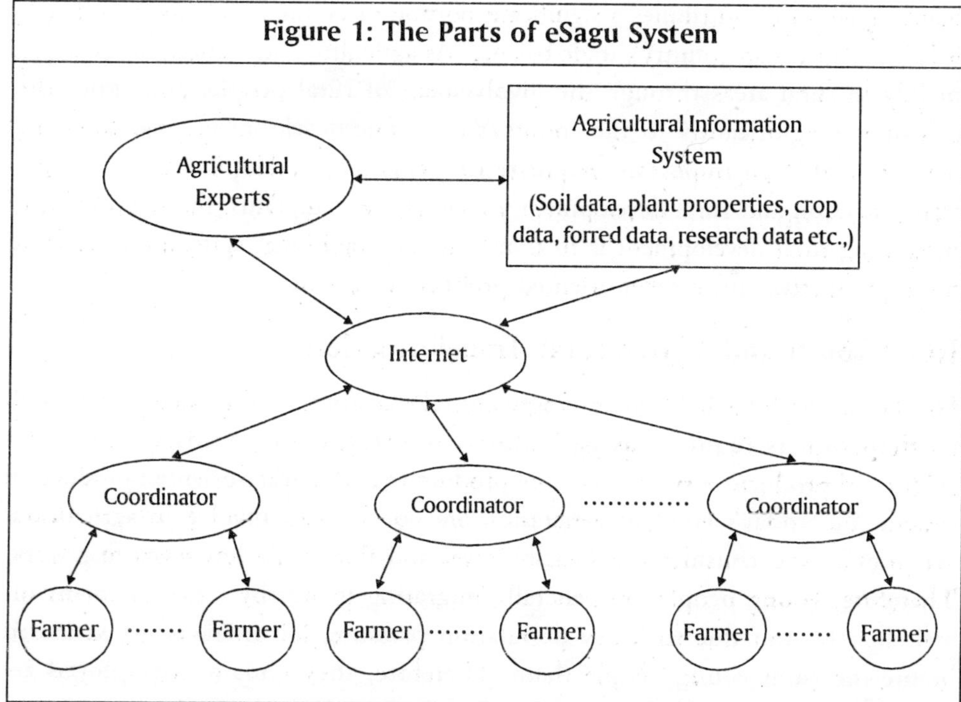

Source: Reddy G Syamasundar and Reddy P Krishna (2006), "Efficient Implementation of Agri-insurance Schemes by Piggybacking eSagu System" Media Lab Asia Project, International Institute of Information Technology, Gachibowli, Hyderabad, India.

[1] Stockholm Challenge 2008, "eSagu: An IT-based personalized agro-advisory system" *http://event.stockholmchallenge.se/project/2008/Economic-Development/eSagu-An-IT-based-personalized-agro-advisory-system*

Agricultural Transformation and Rural Development

Rural Development refers to the structural change in the socioeconomic situation among rural masses so as to improve living standard of low-income people and making the process of their development self-sustaining. In rural areas especially, the source of income is agriculture and therefore the improvement of agricultural sector is the only way to raise the living standard/socioeconomic status of low-income population through structural changes. Agriculture provides a good number of employment opportunities and therefore, it alleviates poverty and promotes food security and nutritional well-being of a nation. Agricultural operation is an activity which uses a nation's major portion of land and water resources. In other words, it plays a crucial role in land and water conservation. This is one of the important processes of maintaining environmental sustainability. On the other hand, agriculture contributes a significant portion of National Income and thereby it helps to maintain country's trade balance. As agricultural operations are practiced mainly in rural areas through the involvement of rural people, and agriculture contributes significantly to the country's economic development, therefore rural development is an important requisite for economic development. Agricultural transformation and rural development are closely related. Thus, the main focus of promoting rural development is to transform the traditional agricultural sector to more productive and market-oriented production system.

Rural Youth and Agricultural Transformation

For the successful achievement of agricultural transformation process, rural youth participation is required as agricultural transformation needs to transform traditional production system to more productive and market-oriented production system. But today's younger generation are not keen to involve in agricultural activities as it is considered a socially lower and financially less rewarding work. Therefore, young people are generally migrating to nearby cities or towns for seeking jobs, but due to heavy population pressure, job markets get saturated before the rural young people them. Therefore, they remain unemployed are vulnerable to anti-social elements. On the other hand, the legal/genuine question that arises is if rural people are not attracted to agricultural operations, then who will cultivate the agricultural land and how will the food problem of the country be solved? The only answer to the question is to make the agricultural production system more productive and market-oriented so that the farmers can sell their product in the internationally competitive markets at high prices which means successful achievement of agricultural transformation. To make agricultural sector more productive and competitive, proper training facilities with modern devices like computer, Internet etc; market information, wide market opportunities, agricultural loan, crop insurance, etc., should be provided to the farmers. These opportunities may attract the rural people to involve in agricultural operations as it may ensure guarantee to earn good income from agricultural sector. To provide such facilities to the farmers, the concerned government's role cannot be overemphasized. Therefore, the government has to be actively involved in providing the required facilities for agricultural operation and at the same time, government should encourage nationalized banks for providing loans, private bank like ICICI Lombard for providing agricultural insurance, the private bodies like ITC (Indian Tobacco Company) and NGOs (Non-Government Organizations), etc., to train the farmers for successful achievement of agricultural transformation.

Agricultural Transformation: A Cross-Country Analysis

Most of the less developed countries in the world, have reported economic stagnation in rural areas, especially, in agricultural sector. Over the last four decades, the agricultural growth was almost stagnant throughout the world though there was considerable increase in Gross National Product (GNP). The decline in per capita food production was the main reason of such stagnation. Since 1970, Africa is experiencing decline in per capita food production as population pressure is increasing. In case of Less Developed Countries (LDCs) where most of the unemployed labor force is engaged in agricultural sector, there occurs low productivity in agricultural sector due to over employment in agriculture. Though this over employment apparently shows that the unemployed labor force is employed, actually, it creates disguised unemployment. Therefore, the development strategy of a nation should include agricultural transformation and thereby achieve rural development.

Agricultural Transformation in Africa

Since 1970, Africa has been experiencing shortage in per capita food production due to increasing population pressure; insufficient supply of inputs for agricultural operation, etc. Water scarcity is one of the main constraints of successful achievement of agricultural transformation. In Africa, total irrigatable land is only 20 per cent of total cultivable land, and only 3 per cent of total available land. The maximum portion of irrigatable land is in Sudan and Gambia. Therefore, most of the African land excluding Sudan and Gambia experiences drought throughout the year. Though few large-scale reservoirs/pumps are found in Africa, most of them failed to supply irrigation facilities for agricultural operations throughout the year.

Soil fertility is another constraint for achieving agricultural transformation. Most of the African land is infertile and thereby generates less yield. Though soil fertility can be improved artificially through the adoption of several techniques like application of fertilizers, soil conservation, biological process such as bioremediation or phytoremediation, etc., but in Africa, adoption of modern technology is very less as compared to the other continents in the world. It is noteworthy to mention that nowadays, the African farmers are gradually beginning to utilize modern techniques of production and are also using chemical fertilizers for the improvement of soil quality. So, is agricultural transformation is happening in Africa? The answer is yes, Africa is coming closer to achieve agricultural transformation. Gambia and Sudan are the examples of observed transformation. Over the last two decades, there has been increased net area under cultivation in the African union and at the same time cropping pattern has been changed from traditional to modern one. "In the past few years, the major stakeholders in African agricultural technology generation and transfer have attempted to reverse the alarming trend of declining per capita agricultural production for the last three decades"[2]. A summary of technology development and transfer is shown in Table 1.

Table 1: Summary of Agricultural Technology Development, Transfer and Commercialization in Africa

Country	Production Technologies	Post-harvest Technologies	NRM Technologies	Total
Kenya	31	3	0	34
Uganda	34	6	0	40
Côte d'Ivoire	14	1	6	21
Mali	9	9	7	25
Ghana	11	9	0	20
Senegal	15	10	0	25
Total	114	38	13	165

[2] Moctar Toure (1996), "Progress and Prospects for Commercialization and Transfer of Agricultural Technologies in Sub-Saharan Africa (SSA): From Maastricht to Accra" *http://www.worldbank.org/html/aftsr/sfi26.doc*

Therefore, it can be stated that Africa is fast moving towards achieving agricultural transformation.

Agricultural Transformation in Latin America

Latin American countries like Chile, Guatemala, Brazil, Peru, etc., have a great potential for the supply of agricultural products in the global market. But the transformation of agricultural sector from traditional to modern is not homogeneous throughout the Latin American countries. It varies from region to region. The benefits of transformation are mainly enjoyed by the modernized processing sector and skillful efficient farmers as this transformation in Latin America has taken place due to the adoption of new agricultural technology. In Latin America around 60 per cent of domestic products are channelized through supermarket to global market. Due to the occurrence of agricultural transformation, rural markets have been diversified into different sectors and rural employment opportunity has been increased. In this continent, agricultural labor market and the rural non-farm economy provides 70 per cent of rural income and 55 per cent of the rural labor force are employed in it[3]. Therefore, agricultural transformation has great contribution towards rural economic prosperity of Latin American countries.

Agricultural Transformation Societies in Spain

As agricultural transformation refers to the process of specializing and making the agricultural produce competitive, it was realized that the achievement of agricultural transformation would be impossible for an individual farmer. For this purpose, Agricultural Transformation Societies were formed throughout the European Union. As a result the European farmers were benefited and the fruits of the Agricultural Transformation were very effective in Spain. Performance of Agricultural Transformation Societies in the Canary Islands of Spain is shown in Table 2.

---

3 Chatterjee Basistha (2008), "Agricultural Transformation in Latin America", published in 'Agricultural Transformation: An Introduction', The Icfai University Press.

Table 2: Structure of the ATS in the Canary Islands according to Productive Activity, 1982-2000

Activity	Societies	%	Members	%	Capital (€)	%
Bananas	81	19.70	4,387	51.35	3,724,073	28.83
Vegetables	69	16.78	334	3.90	1,205,209	9.33
Tomatoes	50	12.16	202	2.36	371,858	2.87
Water	42	10.21	2,042	2.90	1,569,793	12.15
Livestock - cattle	34	8.27	157	1.83	669,813	5.18
Fruits	26	6.32	119	1.39	739,491	5.72
Flowers	23	5.59	143	1.67	1,274,538	9.86
Fruits & Vegetables	21	5.10	104	1.21	590,522	4.57
Winery	17	4.13	480	5.61	1,052,701	8.15
Apiculture	7	1.70	76	0.88	41,356	0.32
Milk & diary products (1)	2	0.48	153	1.79	175,435	1.35
Potatoes	2	0.48	8	0.09	6,912	0.05
Citric fruits	2	0.48	7	0.08	4,207	0.03
Others (2)	8	1.94	36	0.42	119,661	0.92
Total	411	100.00	8,543	100.00	12,915,931	100.00

Notes
(1) In milk and diary products, CELGAN, in Tenerife, has a share capital of 162.273 € and 149 members.
(2) The societies of heliciculture (snails), silk, aviculture, chemical products, seeds and fertilisers, tropical fruits, aromatic herbs, and construction.

Source: Ragistro de Sociedades Agrarias de Transformacion, Consejeria de Agricultura, Pesca y Alimentacion Compilation by the author.

Agricultural Transformation in India

Agricultural transformation in India is significantly visible since 1960s due to the advent of green revolution in the country. During 1960-70 the green revolution took place due to the application of high yielding varieties of seeds, modern technologies, improved irrigation facilities, etc. As a result, India's foodgrain production was increased significantly from 51 million tonnes in 1951-52 to 191.50 million tonnes in 1995-96 to 216 million tonnes 2006-07. On the other hand Gross Domestic Product (GDP) growth continued steadily at 4 per cent during 1992 to 1996 and 2 per cent during 1997-2003 and thereby poverty level was reduced at a significant rate. Table 3 shows trend of foodgrain production during last three decades in India.

However, because of the adoption of technological advancement and other measures towards achieving agricultural transformation, both agricultural productivity and employment generation on agricultural sector have increased throughout the world. Percentage of labor force employed in agricultural sector – 64 in South Asia, 72 in East Asia, 35 in Latin America and 78 in Sub-Saharan Africa and contribution of agriculture as a percentage of Gross National Product (GNP) was 23 in South Asia, 15 in East Asia, 17 in Latin America and 35 in Sub-Saharan Africa during 2006-07.[4]

Table 3: Trend of Foodgrain Production during Last three Decades in India

Year	Production (million tonnes)
1970-71	108
1972-73	95
1978-79	132
1979-80	108
1990-91	176
2001-02	213
2002-03	174
2003-04	212
2006-07	216

Source: Chatterjee Basistha and Chakravarti Visvarup (2008), "Agricultural Transformation in India", The Icfai University Press.

Conclusion

Agricultural transformation refers to the process of specializing and making the agricultural produce towards market-oriented so as to sell the agricultural produce at high prices in the international competitive market. Therefore, it needs structural transformation of agricultural sector through the adoption of several techniques like application of fertilizers, improved technologies, soil conservation, biological process such as bioremediation or phytoremediation, etc. As agricultural transformation ensures to increase agricultural productivity and at the same time helps to produce qualitative and competitive production, therefore, it makes agricultural sector more revenue-generating source. In other words, it brings attention of rural youth to involve more in this sector. More involvement of rural youth in agricultural activities helps on one hand, to solve unemployment problem and on the other hand, solves the food demand of a nation. To make agricultural

[4] Author's compilation from various sources World Development Report and World Development Indicator etc.

sector more productive and competitive, proper training facilities with modern devices like computer, Internet etc; market information, wide market opportunities, agricultural loan, crop insurance, etc., should be provided to the farmers. In this context, concerned Government has to encourage commercial banks, NGOs and other private bodies like ICICI bank, ITC, etc., to facilitate help to the farmers. Government should also encourage agribusiness houses to encourage and practice corporate contract farming.

(Subir Ghosh is Senior Faculty Associate at the Icfai Research Centre, Kolkata. He can be reached at subirg@iupindia.org).

References

Alvaro Balcazar (2003), "Colombian Agricultural Transformation during 1990-2002", *Inter-American Institute for Cooperation on Agriculture, Revista de Economia Institucional*, Vol. 5, No. 9, 2003.

Cándido Roman Cervantes (2006), "Between Solidarity and Profit: The Agricultural Transformation Societies in Spain (1940-2000)", XIV International Economic History Congress, Helsinki 2006, Session 72.

Cervantes Cándido Roman (2006). "Between Solidarity And Profit: The Agricultural Transformation Societies In Spain (1940-2000)", presented in XIV International Economic History Congress, Helsinki 2006, Session 72.

Chatterjee Basistha and Chakravarti Visvarup (2008), "Agricultural Transformation in India".

Chatterjee Basistha (2008), "Agricultural Transformation in Latin America".

Ghimire Krishna (1999), "Rural Youth and Agricultural Transformation in Developing Countries".

Ghosh Subir (2008), "An Introduction to Participatory Rural Development", published in the book *Participatory Rural Development* by The Icfai University Press.

Moctar Toure (1996), "Progress and Prospects for Commercialization and Transfer of Agricultural Technologies in Sub-Saharan Africa (SSA): From Maastricht to Accra".

Proceedings of the seminar on Agricultural Transformation in Africa (1992) held in Baltimore, Maryland May 27-29.

Reddy G. Syamasundar and Reddy P. Krishna (2006), "Efficient Implementation of Agri-insurance Schemes by Piggybacking eSagu System" Media Lab Asia Project, International Institute of Information Technology, Gachibowli, Hyderabad, India.

Stockholm Challenge 2008, "eSagu: An IT-based personalized agro-advisory system".

University of Maryland Eastern Shore (1996), "Commercialization and Transfer of Agricultural Technologies in Sub-Saharan Africa: A Roundtable Workshop", September 4-6, 1996.

http://event.stockholmchallenge.se/project/2008/Economic-Development/eSagu-An-IT-based-personalized-agro-advisory-system

http://www.aec.msu.edu/fs2/ag_transformation/Def_Trans.htm

http://www.lotsofessays.com/viewpaper/1682075.html

http://wps.aw.com/aw_todarosmit_econdevelp_8/0,6111,284734-,00.html

http://www.fao.org/statistics/rural/1.8%20Chapter%20II%20National%20and%20International%20Rural%20Development%20Policies.pdf

http://www.google.co.in/search?hl=e n&q=Between+Solidarity+and+Profit%3A+The+Agricultural+Transformation+Societies+in+Spain&btnG=Google+Search&meta=

2

Farmer Participation in Private Sector Agricultural Extension*

Dominic Glover

Private sector organisations are increasingly important players in the provision of agricultural extension services, as well as crop research, in the global South. It is almost two decades since a group of social science scholars and development practitioners called for a new, participatory approach to agricultural research and extension that would put the 'farmer first' (Chambers et al., 1989). Since then, farmer participatory approaches have been refined, developed, evaluated and critiqued, and can be seen to have influenced the practice and rhetoric of research and extension. This article examines how far these developments have shaped the behaviour of the private sector. Drawing on evidence from a case study, the article argues that the advances of the last quarter-century of thinking and experience with farmer participatory approaches have largely failed to influence the delivery of agricultural extension services to smallholder farmers by a major transnational company.

* This paper is based on Glover, D. (2007) "Farmer participation in private sector agricultural extension", *IDS Bulletin* 38(5), November: 61-73.

Source: www.future-agricultures.org © Dominic Glover. Reprinted with permission. This paper was presented at the Farmer First Revisited Conference at the Institute of Development Studies, University of Sussex, Brighton, UK, 12-14 December 2007.

Introduction

The Monsanto Smallholder Programme (SHP) was an initiative undertaken by the transnational biotechnology, chemicals and seeds company Monsanto between 1999 and 2002. The SHP had the stated purpose of providing 'resource poor', 'smallholder' farmers with a package of agricultural extension services, including technical advice, chemicals, improved seeds and genetically modified (GM) traits, as well as other forms of support.

This article uses the case of the SHP to focus specifically on the issues of farmer participation, responsiveness and accountability, in a situation where extension services were being supplied by a transnational company that is also a major producer and marketer of herbicides as well as the dominant driver behind the development and commercialisation of genetically modified crops internationally. The paper explores how the Smallholder Programme was conceived, designed and implemented, discusses how far it was designed around and responded to farmers' needs and priorities, and considers the extent to which farmers were able to hold the company to account.

Farmer Participation in Agricultural Research and Extension

It is two decades since the first 'Farmer First' workshop took place at IDS and 18 years since the publication of the landmark book with that title (Chambers *et al.*, 1989). The book articulated a critique of the prevailing models of agricultural research and extension and documented a number of cases in which a new, participatory style of engagement with farmers had begun to emerge during the late 1970s and early 1980s. A few years later, *Beyond Farmer First* (Scoones and Thompson 1994) reaffirmed and also extended and developed the farmer first framework, focusing on the power structures that characterised the knowledge systems and institutional contexts in which agricultural research and extension was carried out.

Farmer First and *Beyond Farmer First* were interjections in a much wider stream of discussion and debate. They described, evaluated and critiqued a range of the different approaches that had begun to be experimented with, applied and documented; these included Agro-Ecosystem Analysis, Farming Systems Research and Extension, Farmer Participatory Research, Rapid and Participatory Rural

Appraisal (RRA and PRA), Participatory Action Research and Participatory Technology Development.

Since the early 1990s, a great deal of further experimentation, experience, reflection and debate has occurred. Positions associated with the farmer first school – notably RRA and PRA – have themselves been challenged and criticised, prompting further restatements, adjustments and evolutions such as Participatory Learning and Action (Buhler *et al.*, 2002) and Institutional Learning and Change (ILAC). Another important strand of conceptual development and practical experience has evolved, in a rather separate current, around the Farmer Field School approach (FFS). FFS first emerged in 1989, originally conceived as an educational tool for promoting integrated pest management (IPM) and further developed and promoted by the UN's Food and Agriculture Organisation since then (Braun *et al.*, 2006).

Participants in this conference will be familiar with the kinds of criticisms that were being articulated by the contributors to *Farmer First* and *Beyond Farmer First* almost 20 years ago. The shortcomings and failures they identified were exemplified by the training and visit (T&V) system, espoused by the World Bank, which epitomised the 'transfer of technology' (TOT) mode of engagement. Advocates of a more participatory mode of engagement attacked the TOT approach as being fundamentally inappropriate for the 'complex, diverse and risk-prone' agriculture practised by poor farmers in marginal and rainfed areas in developing countries.

These shortcomings had already been identified, and were increasingly widely recognised, as long ago as the late 1980s. Since then, the concept of a more participatory approach to agricultural research and extension, designed to be more responsive to the needs and priorities of small farmers, has helped to transform the policy debate about how best to organise these services over a period of two decades. And yet, the TOT mode of agricultural research and extension survived, in the T&V system, into the late 1990s, primarily because of the institutional and financial backing of the World Bank (Bauer *et al.*, 1998; Buhler *et al.*, 2002; Anderson *et al.*, 2006). Nevertheless, it was eventually abandoned even by the Bank. However, as I will discuss in this paper, the TOT approach to agricultural research and extension – technology-focused, expert-driven and top-down – survives in the private sector.

The Private Sector in Agricultural Research and Extension

The role of the private sector has attained a new prominence in policy debates about agricultural development in recent years. In search of affordable and effective ways to inject new energy and dynamism into agricultural extension services, various countries have explored policy options including privatisation, greater involvement of private sector service providers, and various models of cost-recovery or fee-charging; India and Pakistan are two examples (Sulaiman and Sadamate 2000; Sulaiman and Hall 2002; Anderson and Feder 2003; Davidson and Ahmad 2003; Shingi *et al.,* 2004). As with the former T&V approach, the World Bank is a key driver of the privatisation agenda (Davidson and Ahmad 2003). In India, a number of agribusiness companies are beginning to explore the market in this area, for example Mahindra & Mahindra's *SubhLabh* Services enterprise (Sulaiman and Hall 2004).[1] But many agricultural input manufacturers and suppliers also claim to offer agricultural extension advice and support alongside the products they sell. However – as staff privately acknowledge – this advice is inflected towards their self-interest in marketing their own brands.

To some extent, private commercial provision of agricultural advice and services implies a degree of responsiveness and accountability on the part of the suppliers towards the customers. However, small and poor farmers are unlikely to exert a strong influence. This implies a limited empowerment of poor rural people to be 'users and choosers' but not necessarily 'makers and shapers' of the goods and services they receive – able to exercise a limited kind of accountability through the market but not much more than that (Cornwall and Gaventa 2001). However, in the case of the Monsanto Smallholder Programme, we see an example of a situation in which a large transnational company geared up a programme specifically targeted towards the small farmer segment of the market. The next section explores how this came about.

Monsanto and Smallholder Farmers

When Robert Shapiro was appointed as Monsanto's new Chief Executive Officer (CEO) in April 1995, he embarked on a programme to re-orient the company's business around 'sustainability'. He linked the urgent need to grow enough food to feed a growing population and the ecological harm that would be entailed in trying to do so with existing technologies and agricultural practices (Scott 1996; Lenzner and Upbin 1997; Magretta 1997; Shapiro 1998; Shapiro 1999). This conceptualisation of the sustainability challenge implicitly involved the farmers and consumers of the developing world as key stakeholders and important players in fulfilling Monsanto's sustainability vision.

Shapiro established seven strategic teams to explore sustainability issues and their implications for the company's future business, including the 'Global Hunger Team' which studied 'how Monsanto might develop and deliver technologies to alleviate world hunger' (Magretta 1997:86). A 'Sustainable Development Business Sector' was established to operationalise the sustainability strategy. It included a 'Smallholder Team' which was 'charged with developing products, services and partnerships to meet the needs of rural, small-scale farmers in developing countries' (Simanis and Hart 2000:A7). Monsanto's agriculture division had already begun to focus on developing country markets in the early 1990s, towards the goal of 'transforming agriculture' in a number of developing countries, a target that became known as the 'developing country goal'.[2] Developing country farmers can thus be seen to have assumed a central place in Monsanto's strategy for market development and competitiveness as well as Shapiro's vision of sustainability.

The concerns of developing countries and smallholder farmers were also among the factors that triggered the crisis which engulfed Monsanto in 1998 and 1999, when the company's plans for the commercialisation of GM crops ran into problems, especially in Europe (Simanis and Hart 2000; Charles 2001; Glover 2007). Both the consumer rejection of GM crops and the specific controversy over so-called 'terminator' sterile seed technology were linked in part to issues of international development and global hunger. For instance, activists and development campaigners raised the alarm over the possibility that terminator technology could make Third World farmers dependent on biotechnology and seed companies (e.g., Christian Aid 1999). Monsanto's advertising campaign provoked an outcry partly because it explicitly asserted the capacity of biotechnology to 'feed the world' (Simanis and Hart 2000; Charles 2001).

While these development controversies helped to trigger the backlash against biotechnology, the image of smallholders and developing country agriculture were also invoked by Monsanto as part of its strategy for tackling the crisis. There were two important reasons for this. With Monsanto's progress in European markets stalled, developing country markets took on greater significance; the company urgently needed to expand the market for its GM crops internationally. In addition, the images of smallholder farmers and poor consumers in developing countries assumed a weighty symbolic importance in global disputes about the merits and risks of GM crops.

The Monsanto Smallholder Programme

Part of Monsanto's response to the crisis it faced in 1999 was to announce the creation of a new Smallholder Programme. The Smallholder Programme was supposed to provide smallholder farmers with 'a package of existing commercial technologies, including improved seeds, biotechnology traits where approved and applicable, conservation tillage practices, crop protection products and other inputs, as well as training and technical assistance'. Monsanto also claimed that the programme provided support for 'self help group formation, ... the creation of other income generating activities [and] access to microcredit, as well as linkages to grain traders and processors who purchase surplus crops' (Monsanto 2002a).

It would be easy to dismiss the SHP as an exercise in public relations, and indeed it is true that Monsanto has used the programme as a source of stories to present the company's relations with Southern farmers in a favourable light. However, the programme was also motivated by a number of other factors. In order to understand these, it is important to recognise the degree to which Monsanto is a relative newcomer to the seed industry, especially in developing countries (Glover 2007).

It was only in the late 1990s that Monsanto acquired interests in a number of international seed companies, including Cargill's global seed business and a stake in the large Indian seed firm, Mahyco. Although Monsanto had been marketing herbicides internationally for many years, its entry into the seed business was a new departure. The SHP in India should therefore be seen in the context of Monsanto's effort to build its market share and develop its brand presence in order to compete with its more established rivals. In addition, the company's global managers in St.Louis were used to dealing with the conditions of large-scale commercial farmers, especially in North America, but Monsanto's strategic expansion in both seeds and herbicides in developing countries brought the company into direct contact with smallholder markets for the first time. The SHP therefore also served as a mechanism to help Monsanto's strategic managers to learn how to engage with the smallholder segment of the market (Glover 2007).

A good summary of the basic SHP model appeared in a Monsanto publication, Growing Partnerships for Food and Health (emphasis added):

> *[D]emonstration plots ar.d farmer trials enable smallholder women and men to witness the value of technology packages that include improved seeds, crop protection products, fertilizers and conservation tillage practices. Training sessions provide the knowledge they need to use the new package safely and effectively. Micro-loans help them get started on their own farms, and market access assistance helps them sell their surplus crops to generate income for their families. Farmers who adopt the new technologies first help expand the effort by teaching others in their community.*
>
> – (Monsanto Undated:4)[3]

In 2001, the SHP was reported to encompass 21 projects in 13 countries, reaching more than 320,000 small farmers (Austin and Barrett 2001), although a company briefing dated January 2002 indicates that Monsanto was directly involved in just a handful of smallholder projects, in Mexico, India, Indonesia, Kenya and South Africa (Monsanto 2002a).[4] In India, the company claimed to be reaching 35,000 farmers in 415 villages.[5] SHP projects were implemented in the states of West Bengal, Madhya Pradesh, Rajasthan and Andhra Pradesh (AP).

The Monsanto *Meekosam* Project

The Monsanto *Meekosam* ('Monsanto for you') project was initiated in Vizianagram District of North Coastal AP in 2001. The project was initiated with a survey of local farmers and agricultural practices in order to identify technology 'adoption gaps'. The project organisers then 'developed a specific package of practices for five or six specific crops' that were deemed to be relevant to the local agricultural systems, including rice, maize, cotton and some horticultural crops. The project promoted particular products and technologies, notably Monsanto's hybrid maize seeds and range of herbicides. It also included some promotional activities for transgenic, insect-resistant Bt cotton.[6]

The project was managed by the local sales manager and supervised by a project co-ordinator, who was a post-graduate in agriculture. Ten Project Officers (POs) were recruited, who were typically recent graduates in agriculture. Each PO was resident in a substantial village and conducted a programme of farm visits, product demonstrations, and farmer training meetings in the surrounding area, as well as responding to farmers' requests for help. The POs maintained checklists to record which farmers had been trained and to track whether they were applying the recommended package of practices.

Farmer Participation in the SHP

In many significant respects, the design of the SHP can be seen to hark back to the T&V style, TOT-mode of agricultural extension. The similarity can be seen, for example, in the rigidity of the uniform package of practices and the assignment of officers to cover a particular catchment area. It is also evident in the use of locally recruited, minimally trained technical assistants in the early projects, who relied on the expertise of more highly trained specialists to deal with farmers' technical queries. Later, as in the *Meekosam* project, graduates were employed as project officers, but nevertheless their activities were closely supervised and organised around the delivery of the approved package of practices through a prescribed schedule of farm visits and other activities. They still relied on technical officers to help them handle more difficult or unusual problems.

Interviews with *Meekosam* project staff make clear that they took for granted a set of assumptions about of information and knowledge transfer, education and training, and the promotion of standardised technology packages to farmers. For instance, the sales manager in charge of the *Meekosam* project in Vizianagram told me, 'Our focus was ... on educating the farmers... They have the need for education and information.... [We were] passing on the technical information – information services to the small growers'.[7] In another interview, he said '[Our purpose was] to help small growers on right cultural practices.'[8] He remarked that the farmers felt the project was 'like a school.'[9] In a similar vein, his counterpart in Warangal District told me that one of the primary objectives of the *Meekosam* Project was 'to give the update of technical information of agriculture as well as our products' and 'our process [was designed] to keep the farmer on track.'[10]

The focus on commercial commodity crops – maize, cotton and rice – resembles what has been labelled the 'commodity approach' to agricultural extension, a mode which harks back to the colonial era. This focus on commodity crops is clearly linked to a set of implicit assumptions about agriculture as a commercial production activity and to Monsanto's aim to promote its major seed and chemical products. These assumptions about commercial farming as an implicit goal for agricultural development were reflected in the way that the SHP was conceptualised by Monsanto executives. For Monsanto, the adoption of new technologies and scientifically informed practices was a central feature of a concept of development through which farmers would make the transition from a subsistence mode of farming to a more commercial agriculture. This technological and commercial transition was in fact central to their image of what 'development' was all about.

'Development' as Technology Transfer and Market Transition

The SHP was conceptualised by Monsanto executives as part of an 'intermediate' or 'transitional' strand of the company's operations that fell between the core business of the firm on one hand, and the company's philanthropic activities (represented by the Monsanto Fund) on the other. This 'three strands' idea appears to have been current among executives at different levels within the Monsanto hierarchy. The concept of the transitional strand carried with it consciously articulated expectations about helping farmers to make the leap from one realm to the other – as one senior SHP executive put it, 'from the subsistence to the commercial world', 'from subsistence to market'.[11]

The idea of development implicit in the concept of the transitional strand is encapsulated in the notion of farmers making a transition from non-commercial (perhaps 'pre-commercial') subsistence farming to commercial agriculture. The key event which demonstrates this transition is the farmers' adoption of new technology. According to this conception, non-adoption or slow adoption of new technology is seen as a key signifier of underdevelopment, and the act of technology adoption signifies development taking place. At the same time, technology adoption represents the farmers' transition to a new, commercial mode of agriculture, in which the purchase of external inputs is a key feature. Hence, in this conception poor farmers assumed a dual identity as both potential beneficiaries and potential customers of Monsanto's technology. But were smallholders the primary beneficiaries of the SHP projects?

Targeting Smallholders?

Monsanto's publications emphasise the company's engagement with 'smallholders'. However, it is not clear that smallholders, strictly defined, were the primary target group. The company appeared to define smallholders as those farming less than five hectares of land (12.4 acres) (Monsanto Undated:4) – which is not particularly small; in India, small and marginal farmers are commonly defined as those cultivating areas around one acre.[12] In fact, however, it is not clear that the size of a farmer's landholding or other concrete indicators were used to select participants for the smallholder projects. A St.Louis-based Monsanto marketing executive told me 'I'm not sure we had a real clean definition' of smallholders, noting that the meaning of the concept would vary from place to place. He preferred to think in terms of a rule of thumb which took into account the 'size' and 'economics' of the farm but also, crucially, its 'long-term potential' and the 'objectives of the farmer'.[13]

This reference to the 'objectives of the farmer' is a code for the farmer's willingness to adopt new technologies and manage his or her farm on a commercial footing. Farmers who were thought to be most likely to experiment with new technologies were characterised as 'leading' or 'progressive' farmers, whereas smallholders were characterised as 'slow adopters' of new technology.[14] However, the underlying issue of the differences among farmers in terms of their capacities to experiment with technology – to afford it, in particular – was not generally acknowledged – although one senior marketing executive in St.Louis did recognise that the poorest farmers would 'never' adopt Monsanto's GM crop technologies, 'because they just can't afford the up-front costs'.[15]

In practice, both small and large farmers participated in SHP projects, but larger and more prosperous farmers were given special attention. As one observer of the project in Kumili village pointed out, the project officials had provided free samples of Monsanto's maize hybrids 'not for the weaker sections, only for the gentlemen of the village... That way, the gentlemen will explain to the common man.'[16] The need to prioritise larger and more prosperous 'lead farmers' or 'key important farmers',[17] in the expectation that their example would influence other farmers in the project villages, was taken for granted by Monsanto staff at all levels of the company.

The emphasis on lead farmers calls into question the degree to which the SHP represented a novel departure from agribusiness companies' regular market development activities, which would typically involve much of the same basic repertoire of farmer meetings, demonstration plots, engagement with opinion leaders and so on. Indeed, the key difference between the SHP and regular marketing was not so much in the techniques used but the market segments targeted. As Monsanto's former global manager for smallholder agriculture observed, the commercial operations staff 'were dealing with the low-hanging fruit; we were dealing more with the second tier'.[18]

In this respect, the SHP projects arguably have a good claim to additionality, in that they were serving areas of the market that wouldn't ordinarily have been reached. However, by targeting the larger farmers in such districts, it has to be doubted whether the projects were really reaching the lower levels of the market. As if to illustrate this point, one apparently prosperous farmer who had participated in the Guntur pilot project told me that the project had not made much difference to him individually, since he had always been able to access the inputs he needed from his local distributor, on credit, both before and after the project.[19]

In certain respects, the SHP can be seen to have been a novel initiative that was in some degree responsive to the needs of farmers. A key factor was the assignment of project officers to live in villages. This meant that they were available on the spot to deal with questions and problems brought to them by farmers. According to one farmer who lived close to the *Meekosam* project office in Kumili village, a queue of farmers would form outside the office every morning.[20]

Project officers reported that they spent a significant proportion of their time responding to the farmers' requests for help and advice.

The fact that the POs were resident as guests in the village, combined with their professional responsibility to promote the Monsanto brand as a dependable and approachable source of technical expertise, placed them under a strong obligation to help the farmers resolve all kinds of agronomic problems, even where that didn't involve an opportunity to promote Monsanto's own products.[21] Indeed, at least in the early phase of the SHP, senior managers strongly emphasised that the programme was not intended primarily to promote Monsanto products; staff were told "Don't link it with commercial [operations]" and 'not a single word was uttered on Monsanto products'.[22]

To some degree, therefore, the farmers were able to exercise an immediate kind of influence over the content and delivery of the programme, and a face-to-face form of accountability. Indeed, farmers in Vizianagram District credited Monsanto with offering them services which no other agency was providing. They held the *Meekosam* project officers in high esteem and expressed gratitude for the work they did. So what was the role of farmers in the SHP?

The Appearance of Farmers

The Monsanto managers' descriptions of the SHP as an information and training programme, to 'educate the farmers' and 'keep them on track', depicted small farmers as uninformed, slow to adopt new practices and technologies, in need of education to improve their farming methods and wayward in their failure to adhere to the practices they were taught unless they were closely supervised. In this implicit conception, the small farmers appear as rather passive recipients of new knowledge. In fact, however, the farmers themselves asserted that they were willing to experiment with new technologies to see if they lived up to the promised benefits.

For example, one farmer in Karlapudi village, Guntur, announced that 'whatever's new, I'll go for it – seeds, pesticides, whatever.'[23] But he was equally willing to abandon the new practices if they didn't work or he found them unsuitable. He related how the project officers had strongly encouraged farmers to adopt new pest control techniques, namely stem application of pesticides and the planting of castor and marigold as 'trap crops'. Although he had tried the techniques during the project, he abandoned them immediately afterwards because they were labour-intensive and 'we're already used to spraying.... Labour is also a problem here – that's why we're not going for trap crops'.[24] The same farmer smiled as he acknowledged that he and his neighbours had effectively ignored the advice they were given: 'that particular year they were very particular about not using pesticides indiscriminately [but] once the project was over, we reverted to the normal [indiscriminate] spraying'.[25]

One of his neighbours said that he had abandoned the recommended practices because he felt they didn't work well enough: 'For the first one hundred days, we followed their advice but the pest was not under control.... We adopted our own methods after that'.[26] I asked whether he had continued with the use trap crops as part of his pest management system: 'No. If we plant marigolds, it attracts porcupines, and that destroys our crop'.[27] In another village, the farmers abandoned some of the recommended products and techniques when the *Meekosam* project ended, primarily because support was no longer so readily available on their doorstep: 'Earlier, we used to get the samples here, now we have to go to Vizianagram or Pusapatirega to get advice'.[28]

These comments help to show the degree to which the inflexible, expert-defined packages of practices failed to address the needs and priorities of the participating farmers in a coherent and sustainable way. There were no meaningful opportunities for farmers to express their views, determine the priorities of the projects or contribute to the design of the technology packages. When the new practices were found to be unsuitable, farmers were able to exercise their choice as consumers to reject or abandon them, but there were no meaningful channels through which their feedback might have been used to reshape the projects.

More fundamentally, there were no opportunities for the farmers to influence Monsanto's upstream research and development priorities. Indeed, one senior executive attested that he didn't see much opportunity for small farmers to help shape the corporate R&D programme, because the strategic decision to pursue a particular crop or trait technology was a 'high-level' one, early in the innovation process, after which there was 'not much fine tuning' to which farmer feedback might contribute.[29]

The model of agricultural extension embodied in the SHP was essentially a commercial one, informed by assumptions based on a consumerist, market-based model of technology diffusion. Indeed, the focus on market development was entrenched in the organisation of the SHP. At all levels of the programme, the small SHP team depended rather heavily on the cooperation of their colleagues in sales and market development to help implement their projects. Sales and market-research data were used to identify potential areas for SHP projects, and the final choice of particular villages was influenced by local sales staff. As one senior SHP executive put it, 'we were guided by our marketers' to areas 'where they thought we could contribute the most'.[30] Sales staff were also directly involved with the day-to-day management and supervision of SHP projects.

Over time, the influence of sales and market development imperatives had a strong influence on the content and implementation of the programme. SHP staff found that they had to continually explain and 'sell' the programme to operational colleagues, who were sometimes sceptical or even hostile. They increasingly did so by justifying the programme as a means of developing experimental 'models' for future marketing efforts to the smallholder sector. Over time, this undoubtedly shaped perceptions of what the programme was for. As a result, as time passed, business imperatives came to dominate over the more philanthropic goals of the programme. The emphasis on commercial goals ultimately undermined the perceived rationale for a special programme targeted towards smallholders, which helps to explain why the programme was terminated, prematurely, in 2002 (Glover 2007).

Conclusion

The Monsanto Smallholder Programme can be seen to have evoked outmoded and largely discredited approaches to agricultural research and extension, closely resembling the T&V system. The programme was designed and implemented in a top-down, expert-driven mode which aimed to facilitate a one-way transfer of technology from Monsanto's laboratory scientists and plant breeders to farmers. The company drew on the 'expertise' of scientists to design a standard package of practices for each focus crop, constructed around the adoption of particular technologies, assumed to be universally applicable and intended to be delivered in a uniform manner across a given territory. The selection of focus crops themselves was determined primarily by the products and technologies Monsanto had to offer and wished to promote, and the selection of districts and villages in which to implement projects was shaped by the market development priorities identified by sales managers.

The need for farmers to adopt new technologies and commercial approaches to farming were regarded as axiomatic steps in the process of 'development', so that, instead of designing the SHP projects in response to or with input from the farmers themselves, their role in Monsanto's implicit vision of development was essentially a passive one: to be consumers of the company's know-how and technological products, and to take their place in Monsanto's scheme for sustainable agriculture in the twenty-first century – the vision of 'transforming agriculture in developing countries'.

Part of the reason for this is that Monsanto needed to generate favourable associations between GM technology and small farmers. Smallholders had assumed a particular symbolic importance in global arguments about transgenic crops. Monsanto has sought to document and publicise its work with 'smallholders' to a wide audience, through a series of publications and websites (Monsanto 2001; Monsanto 2002b; Monsanto 2002a; Monsanto 2004; Monsanto 2005; Monsanto Undated).[31] By highlighting positive stories about the benefits of GM technology for smallholders, Monsanto clearly hopes to influence the global debate about GM crops, and in particular to demonstrate or even 'prove' to the wider world the value and appropriateness of transgenic crops for developing countries. In order for this argument to be made, it was necessary to make certain prior assumptions about the needs and priorities of farmers, who were constructed as rather passive recipients of new agricultural technology.

Whereas Monsanto's staff regarded the adoption of the new methods and practices as a one-time, one-directional developmental step that would transform farmers' behaviour into the future, farmers themselves were willing to experiment with new techniques but were happy to abandon them and go back to their previous practices if, for whatever reason, they found them undesirable – for example, ineffective, inconvenient or expensive. To that extent, they can be seen to have exercised voice as consumers in the market-place, but a greater opportunity to design a programme around their expressed needs and priorities was lost.

(Dominic Glover Post-doctoral fellow, ESRC STEPS Centre, IDS, Sussex. The author can be reached at dominic.glover@wur.nl).

Endnotes

1 See *http://www.mahindra.com/mahindras/FARM_EQUIPMENt/mssl/mssl.htm#MSSL* (19 May 2006).

2 Interview, Monsanto executive, St.Louis, 20/06/05. My source could not recall the exact terms of the developing country goal, including the number of countries targeted for agricultural 'transformation'.

3 The document itself is undated. However, two inserts, tucked into a pocket in the flyleaf at the back of the document, are dated January 2002. The packet was given to the author by a Monsanto executive in St.Louis in May 2002.

4 According to the same document, Monsanto was indirectly involved in projects in a larger number of countries, through its support for projects implemented by Winrock International in 'West Africa and Indonesia' and by Sasakawa Global 2000 in 'Ghana, Ethiopia, Tanzania, Malawi and Mozambique'. A former 'Global Lead, Smallholder Agriculture' described this arrangement as 'more like aid' (Interview, former Monsanto executive, USA, 29/06/05).

5 *http://www.monsantoindia.com/monsantoin/humsafar/humsafar2.html* (21/05/04).

6 Information compiled from interviews with current and former Monsanto India managers, *Meekosam* project staff and local farmers between October 2004 and December 2005.

7 Interview, former Monsanto India sales manager, Hyderabad, 14/11/05.

8 Interview, former Monsanto India sales manager, Hyderabad, 22/11/05.

9 Interview, former Monsanto India sales manager, Hyderabad, 24/11/05.

10 Interview, former Monsanto India sales manager, Vijayawada, 03/12/05.

11 Interview, former Monsanto executive, USA, 29/06/05.

12 From St.Louis, of course, almost all Indian farmers would be considered very small.

13 Interview, Monsanto executive, St.Louis, 23/06/05.

14 Multiple interviews, e.g. former *Meekosam* PO, Guntur, 4/12/05; former Monsanto India sales manager, Hyderabad, 24/11/05; Monsanto India SHP executive, Mumbai, 11/01/05; Monsanto India SHP executive, Mumbai, 30/03/05.

15 Interview, Monsanto executive, St.Louis, 27/06/05.

16 Interview, retired teacher, Kumili village, Vizianagram, AP, 16/12/05.

17 Interview, former *Meekosam* PO, Guntur, AP, 04/12/05.

18 Interview, former Monsanto executive, USA, 29/06/05.

19 Interview, farmer, Karlapudi village, Guntur, AP, 15/12/05.

20 Interview, farmer, Kumili village, Vizianagram, AP, 16/12/05.

21 Interview, Monsanto India executive, Delhi, 18/11/04.

22 Interview, Monsanto India executive, Mumbai, 11/01/05.

23 Interview, farmer, Karlapudi village, Guntur, AP, 15/12/05.

24 Ibid.

25 Ibid.

26 Interview, farmer, Karlapudi village, Guntur, AP, 15/12/05.

27 Ibid.

28 Interview, farmer, Kumili village, Vizianagram, AP, 16/12/05.

29 Interview, Monsanto executive, USA, 20/06/05.

30 Interview, former Monsanto executive, USA, 29/06/05.

31 Monsanto's Pledge website has a section on the socioeconomic impacts of GM crops (*http://www.monsanto.com/monsanto/layout/our_pledge/socioeconomic*, 23/02/06), which complements special websites that promote biotechnology such as The Biotech Knowledge Center (*http://www.biotechknowledge.monsanto.com*, 6/09/06) and 'Conversations about Biotechnology' (*http://www.monsanto.com/biotech-gmo/index.htm*, 6/09/06). In 2000, Monsanto launched a specific programme to encourage and facilitate, directly and indirectly, research by independent academics into the economic and social impacts of genetically modified, insect-resistant Bt cotton (Monsanto 2001).

References

Anderson, J. R. and G. Feder (2003) "Rural Extension Services", IBRD Policy Research Working Paper, Washington, DC, USA: The World Bank, Agriculture and Rural Development Department and Development Research Group.

Anderson, J. R., G. Feder and S. Ganguly (2006) "The Rise and Fall of Training and Visit Extension: An Asian Mini-Drama with an African Epilogue", IBRD Policy Research Working Paper 3928, Washington, DC, USA: The World Bank, Development Research Group, Rural Development Team.

Austin, J. E. and D. B. Barrett (2001) "Monsanto. Technology Cooperation and Small Holder Farmer Projects", Harvard Business School case study, Cambridge, MA, USA: Harvard Business School.

Bauer, E., V. Hoffmann and P. Keller (1998) "Agricultural extension down the ages", Agriculture + Rural Development 5(1), April: 3-6.

Braun, A., J. Jiggins, N. Röling, H. van den Berg and P. Snijders (2006) "A Global Survey and Review of Farmer Field School Experiences", Wageningen, NL: Endelea.

Buhler, W., S. Morse, E. Arthur, S. Bolton and J. Mann (2002) *Science, Agriculture and Research: A Compromised Participation*, London: Earthscan.

Chambers, R., A. Pacey and L.-A. Thrupp, eds. (1989) *Farmer First: Farmer innovation and agricultural research*, London: Intermediate Technology Publications.

Charles, D. (2001) *Lords of the Harvest: Biotech, big money, and the future of food*, Cambridge, MA, USA: Perseus.

Christian Aid (1999) "Selling Suicide. Farming, false promises and genetic engineering in developing countries", London: Christian Aid.

Cornwall, A. and J. Gaventa (2001) "From users and choosers to makers and shapers: repositioning participation in social policy", IDS Working Paper 127, Brighton: Institute of Development Studies.

Davidson, A. and M. Ahmad (2003) *Privatization and the Crisis of Agricultural Extension: The Case of Pakistan*, Aldershot, UK: Ashgate.

Glover, D. (2007) "Monsanto and smallholder farmers: a case study in corporate social responsibility", *Third World Quarterly* 28(4): 851-867.

Lenzner, R. and B. Upbin (1997) "Monsanto v. Malthus", *Forbes*.

Magretta, J. (1997) "Growth Through Sustainability: An Interview with Monsanto's CEO, Robert B. Shapiro", *Harvard Business Review*: 79-88.

Monsanto (2000) "Monsanto Chief Executive Outlines Commitments on New Agricultural Technologies in the 'New Monsanto Pledge", St.Louis, Missouri: Monsanto. www.monsanto.com

Monsanto (2001) "Fulfilling Our Pledge: 2000–2001 Report", St.Louis, Missouri: Monsanto.

Monsanto (2002a) "Backgrounder: Monsanto Small Holder Program", January.

Monsanto (2002b) "Commitments to Our Stakeholders: 2001–2002 Monsanto Pledge Report", St.Louis, Missouri: Monsanto.

Monsanto (2004) "Growing Options: Monsanto Company 2004 Pledge Report", St.Louis, Missouri: Monsanto.

Monsanto (2005) "Seeding Values. 2005 Pledge Report", St.Louis, Missouri: Monsanto.

Monsanto (Undated) "Growing Partnerships for Food and Health: Developing Country Initiatives in Agricultural Product and Technology Cooperation".

Scoones, I. and J. Thompson, (eds.) (1994) *Beyond Farmer First: Rural people's knowledge, agricultural research and extension practice*, London: Intermediate Technology Publications.

Scott, M. (1996) Interview: Robert Shapiro, CEO of Monsanto, Co.: Monsanto's Brave New World, Business Ethics Magazine.

Shapiro, R. B. (1998) "Trade, Feeding the World's People and Sustainability: A Cause for Concern", The CEO Series 22, St.Louis, MO, USA: Center for the Study of American Business, Washington University in St.Louis.

Shapiro, R. B. (1999) "How genetic engineering will save our planet", *The Futurist*. 33: 28-29.

Shingi, P. M., V. R. Gaikwad, R. Sulaiman and J. Vasanthakumar (2004) Agricultural Extension, New Delhi: Academic Foundation.

Simanis, E. and S. Hart (2000) "The Monsanto Company: Quest for Sustainability", Washington, DC: World Resources Institute, Management Institute for Environment and Business.

Sohoni, S., M. S. Mithyantha and P. Ravi (2000) "Rallis Chaitanya Farm Management Project", *Kisan World* 27(10), October: 46-47.

Sulaiman, R. and A. Hall (2002) "Beyond Technology Dissemination – Can Indian agricultural extension reinvent itself?" NCAP Policy Briefing 16, New Delhi: National Centre for Agricultural Extension Policy and Research.

Sulaiman, R. and A. Hall (2004) "Towards Extension-plus: Opportunities and Challenges", NCAP Policy Brief 17, New Delhi: National Centre for Agriculturaly Economics and Policy Research.

Sulaiman, R. and V. V. Sadamate (2000) "Privatising Agricultural Extension in India", NCAP [National Centre for Agricultural Economics and Policy Research] Policy Paper (10).

3

Effects of Agricultural Management on Soil Organic Matter and Carbon Transformation – A Review

Xiaobing Liu, S J Herbert, A M Hashemi, X Zhang and G Ding

Soil Organic Carbon (SOC) is the most often reported attribute and is chosen as the most important indicator of soil quality and agricultural sustainability. In this review, we summarized how cultivation, crop rotation, residue and tillage management, fertilization and monoculture affect soil quality, Soil Organic Matter (SOM) and carbon transformation. The results confirm that SOM is not only a source of carbon but also a sink for carbon sequestration. Cultivation and tillage can reduce soil SOC content and lead to soil deterioration. Tillage practices have a major effect on distribution of C and N and the rates of organic matter decomposition and N mineralization. Proper adoption of crop rotation can increase or maintain the quantity and quality of soil organic matter, and improve soil chemical and physical properties. Adequate application of fertilizers combined with farmyard manure could increase soil nutrients, and SOC content. Manure or crop residue alone may not be adequate to maintain SOC

Source: Plant Soil Environ, 52, 2006 (12): 531-543 (http://www.cazv.cz).

levels. Crop types influence SOC and soil function in continuous monoculture systems. SOC can be best preserved by rotation with reduced tillage frequency and with additions of chemical fertilizers and manure.Knowledge and assessment of changes (positive or negative) in SOC status with time is still needed to evaluate the impact of different management practices.

Soil is a vital natural resource that is nonrenewable on the human time scale (Jenny 1980) and is a living, dynamic, natural body that plays many key roles in terrestrial ecosystems. It is the essence of life and health for the well-being of humankind and animals and the major source of most of our food production. The maintenance of soil health is essential for sustained productivity of food, the decomposition of wastes, storage of heat, sequestration of carbon, and the exchange of gases. However, only a limited area of the soil can actually be used for growing food, and when improperly managed it can be eroded, polluted or even destroyed (Brady and Weil 2000).

It is estimated that the soil in 11% of the vegetative area and 38% of the cultivated area in the world have been degraded since 1945 (Hammond 1992, Gardiner and Miller 2004). This is an area of the size of China and India together. Approximately 24 billion tons of topsoil is lost annually, which is equivalent to about 9.6 million hectares of land (Bakker 1990). Therefore, soil degradation and/or changes in soil quality that result from wind and water erosion, salinization, losses of organic matter and nutrients, or soil compaction are of great concern in every agricultural region in the world.

In the last two decades, public interest in soil quality has been increasing throughout the world as humankind recognizes the fragility of earth's soil, water and air resources, and the need of their protection to sustain civilization. The concept of soil quality was first suggested in 1977 at a conference (Doran and Parkin 1994) which focused on the risks and benefits associated with intensive agriculture, but the concept *per se* was not discussed until 1980s when it was defined based on the soil function, and the methods to evaluate it were published.

Soil quality was defined in many different ways. Power and Meyers (1989) defined soil quality as the ability of soil to support crop growth, including factors such as tilth, aggregation, organic matter content, soil depth, water holding capacity, infiltration rate, pH changes and nutrient capacity. Larson and Pierce (1991) defined soil quality as the capacity of the soil to function within ecosystem boundaries and to interact positively with the environment external to that ecosystem. After the US National Academy of Science published "Soil and Water Quality: An Agenda for Agriculture" in 1993, the concept evolved with a holistic focus emphasizing that sustainable soil management required more than soil erosion control. Mausbach and Tugel (1995) similar to Larson and Pierce (1991) stated that "soil quality reflects the capacity of a specific kind of soil to function within natural or managed ecosystem boundaries, to sustain plant and animal productivity, maintain or enhance water and air quality, and support human health and habitation". Sojka and Upchurch (1999) further proposed that soil quality must be defined in terms of distinct management and environmental considerations specific to one soil, under explicit circumstances for a given use. The considerations included social, economic, biological and other value judgments. Singer and Ewing (2000) stated that "useful evaluation of soil quality requires agreement about why soil quality is important, how it is defined, how it should be measured, and how to respond to measurements with management, restoration, or conservation practices". Thus, these issues are not easily addressed because determining soil quality requires one or more value judgments and because there is still much unknown about soil.

Worldwide research and technology transfer efforts have increased awareness that soil resources have both inherent characteristics determined by their formation factors and dynamic characteristics determined by human decisions and management practices. In this sense, understanding soil quality means assessing and managing soil so that it functions optimally now and is not degraded for future use (Brady and Weil 2002). Much like air and water, the quality of soil has a profound effect on the health and productivity of a given ecosystem and the environments related to it. However, unlike air and water for which we have quality standards, the definition of soil quality is complicated because in part humans and animals do not directly consume it (Doran and Parkin 1994, Liu and Herbert 2002). Soil quality cannot be measured directly, so there is a need to evaluate indicators. Soil indicators are measurable properties of soil or plants that provide clues about how well the soil can function. A variety of physical, chemical, and biological characteristics can be used as indicators. However, descriptive indicators and quantitative means to monitor soil quality are, at best, difficult to define. Arshad and Coen (1992) gave possible descriptive indicators to characterize soil quality, which included evidence of erosion, soil structure, friability, crusting of the soil surface and ponding of water. All of these are physical attributes of the soil, to a certain extent they can be controlled by Best Management Practices (BMPs), and they are somewhat qualitative in nature. Quantitative measures to monitor soils such as soil pH and extractable N-P-K are more developed but are still being explored as to how these measures affect yield, nutrient levels and the biological health of the soil. Soil quality indicators interact with one another, and thus the value of one is affected by one or more of the other selected parameters. Soil quality also varies due to many external factors such as land use, soil and crop management, environmental interactions, societal goals as well as variation in natural conditions (Campbell *et al.,* 2001a, Follett 2001, Arshad and Martin 2002, Dao *et al.,* 2002, Liu *et al.,* 2003, Liebig *et al.,* 2004).

Soil Organic Matter (SOM) is the central indicator of soil quality and health, which is strongly affected by agricultural management (Lal *et al.,* 1995, Farquharson *et al.,* 2003). SOM is a major terrestrial pool for C, N, P, and S, and the cycling and availability of these elements are constantly being changed by microbial immobilization and mineralization (Hillel 1991, Feichtinger *et al.,* 2004). The importance of increased SOM or Soil Organic carbon (SOC) is its effect on improving soil physical properties, conserving water, and increasing available nutrients. These improvements should ultimately lead to greater biomass and crop yield (Bauer and Black 1994, Berzsenyi *et al.,* 2000, Onemli 2004). There is considerable concern that if SOM or SOC concentrations in soils are allowed to decrease too much, the productive capacity of agriculture will be then compromised by deterioration in soil physical properties and by impairment of soil nutrient cycling mechanisms Bauer and Black 1994, Loveland and Webb 2003). The interactions and negative impacts of agricultural management on soil quality are illustrated in Figure 1.

Figure 1: Diagram of Interactions and Negative Impacts of Agricultural Management on Soil Quality

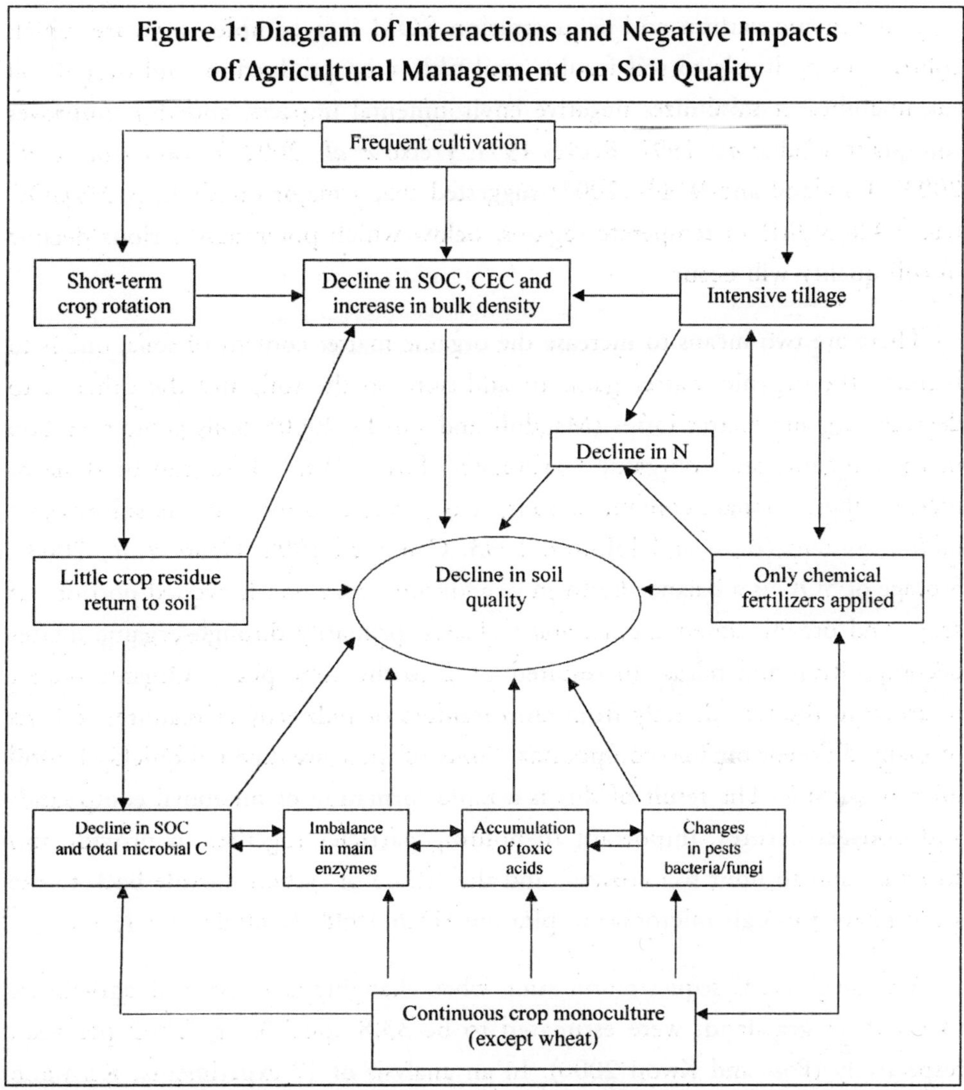

Long-term experiments are often required to predict soil management impacts on soil carbon storage and provide leading indicators of sustainability, which can serve as an early warning system to detect impairments that threaten future productivity (Clapp *et al.*, 2000).

In this review, we summarized how cultivation, crop rotation, residue and tillage management, fertilization, and monoculture affect soil quality, SOM and carbon transformation.

Soil organic carbon and sequestration SOM is a complex mixture which contributes positively to soil fertility, soil tilth, crop production, and overall soil sustainability. It minimizes negative environmental impacts, and thus improves soil quality (Lal *et al.*, 1997, Reeves 1997, Freixo *et al.*, 2002, Farquharson *et al.*, 2003). Loveland and Webb (2003) suggested that a major threshold is 2% SOC (ca. 3.4% SOM) in temperate regions, below which potentially serious decline in soil quality will occur.

There are two means to increase the organic matter content of soils; one is to increase the organic matter gains or additions to the soil, and the other is to decrease organic matter losses (Magdoff and van Es 2000). Soils contain carbon in their organic and inorganic (carbonates) forms. While little can be done to increase the carbonate content of soils, it is possible to increase the soil organic matter content (Kern and Johnson 1993, Campbell 1998, Chan *et al.*, 2003). Storage of SOC is a balance between C additions from non-harvested portions of crops and organic amendments, and C losses, primarily through organic matter decomposition and release of respired CO2 to the atmosphere. Organic matter returned to the soil, directly from crop residues or indirectly as manure, consists of many different organic compounds. Some of these are digested quickly by soil microorganisms. The result of this is a rapid formation of microbial compounds and body structures, important in holding particles together to provide soil structure and to limit soil erosion, and the release of carbon dioxide back to the atmosphere through microbial respiration (Dao 1998, Kladivko 2001).

Average global C sequestration rates, when changing land use from agriculture to forest or grassland, were estimated to be 33.8 and 33.2 g C/m2 per year, respectively (Post and Kwon 2000). In an analysis of 17 experiments, Kern and Johnson (1993) concluded that the greatest change of C occurred in the top 8 cm of soil, a lesser amount in the 8- to 15-cm depth, and no significant amount below 15 cm. They also concluded that, unlike no-till (NT) in which SOC had increased, no significant change in SOC was detected in response to Reduced Tillage (RT). From these studies, they estimated the duration of C sequestration to be between 10 and 20 years. Paustian *et al.*, (1998) compared tillage systems, ranging in duration from 5 to 20 years, and estimated that NT resulted in an average soil C increase of 285 g/m2, compared to Conventional Tillage (CT). Using an average experiment duration of 13 years it implies an approximate C sequestration rate of 22 g/m2/yr. West and Post (2002) analyzed the C sequestration rates using a global database of 67 long-term agricultural experiments and indicated that on average, a change from CT to NT can sequester 57 ± 14 g C/m2/yr, and by enhancing crop rotation complexity it can sequester an average of 20 ± 12 g C/m2/yr. Carbon sequestration rates can be expected to peak in 5–10 years with SOC reaching a new equilibrium in 15–20 years. Following initiation of an enhancement in rotation complexity, SOC may reach a new equilibrium in approximately 40–60 years. The analysis of long-term experiments in Canada indicated that SOC can be sequestrated for 25–30 years at a rate of 50–75 g C/m2/yr, depending on soil type, in well fertilized Chernozemic and Luvisolic soils cropped continuously to cereals and hay (Dumanski *et al.,* 1998).

For total SOC calculation and global carbon cycle, carbon content is often expressed on an area-basis (depth-based) for comparisons (Ding *et al.,* 2002). Liu *et al.,* (2003) showed a significant decline of total SOC that occurred in the first 5 years of cultivation where the average SOC loss per year was about 2300 kg/ha for the 0–17 cm horizon. The average annual SOC loss between 5- and 14-year cultivation was 950 kg/ha and between 14- and 50-year cultivation it was 290 kg/ha. These data clearly showed a rapid reduction of SOC for the initial soil disturbance by cultivation and a relatively gradual loss later. Compared with organic matter in the uncultivated soil, Liu *et al.,* (2003) also indicated that the total SOC loss (the sum of three horizons) was 17%, 28%, and 55% in the 5-, 14- and 50-year cultivation periods, respectively. The latter would correspond to the release of approximately 380 ton CO2/ha to the atmosphere. These results were consistent with the estimation of changes in SOC during cultivation by Davidson and Ackerman (1993). This means that a new equilibrium will be reached when SOC declines to a threshold.

Thus, soil organic matter is not only an important source of carbon for soil processes but also a sink for carbon sequestration. Cultivation can reduce SOC content and lead to soil deterioration, and finally reduce soil productivity. By changing land use and tillage systems or by the adoption of sustainable crop rotations and the inclusion of perennial vegetation, carbon sequestration rates can be increased to a range of 20–75 g/m2/yr, and SOC may reach a new equilibrium within several decades.

Soil Tillage

Tillage is used to mix and aerate the soil, and to incorporate cover crops, crop residue, manure, fertilizers and pesticides into the rooting zone (Acquaah 2002). Soil tillage management can affect factors controlling soil respiration, including substrate availability, soil temperature, water content, pH, oxidation-reduction potential, kind and number of microorganisms, and the soil ecology (Robinson *et al.,* 1994, Kladivko 2001).

Beare *et al.,* (1994) indicated that tillage enhanced short-term CO_2 evolution and microbial biomass turnover, and accelerated organic C oxidation to CO_2 not only by improving soil aeration, but also by increasing contact between soil and crop residues, and by exposing aggregate-protected organic matter to microbial attack. Tillage also exposes organic carbon in both the inter- and intra-aggregate zones and that immobilized in microbial cellular tissues to rapid oxidation (Roscoe and Burman 2003). This is because of the improved availability of O_2 and the exposure of more decomposition surfaces, thereby stimulating increased microbial activity (Beare *et al.,* 1994, Jastrow *et al.,* 1996).

Conventional tillage significantly reduces biological diversity in surface soil. Doran and Smith (1987) summarized data from the USA, Canada, and England and stated that the surface level of SOC, microbial biomass, and potentially mineralizable N was all significantly higher in the no-till than in the plow-till method of seedbed preparation.

Increased C storage has been frequently observed in soils under conservation tillage, particularly with NT (Unger 1991, Zibilske *et al.,* 2002). A widespread adoption of conservation tillage could result in net increases in C sequestration in agricultural lands, reversing the decline caused by intensive tillage practices used for decades (Kern and Johnson 1993, Campbell *et al.,* 2001b). Kushwaha *et al.,* (2001) found that the values of SOC and total N were the highest in the minimum tillage and residue retained treatment and the lowest in conventional tillage and residue removed treatment. Tillage reduction from conventional to minimum and zero conditions along with residue retention increased the proportion of macro-aggregates over 21-42% over control in soil (Kushwaha *et al.,* 2001). Active microbial biomass and C mineralization were higher under NT than under conventional tillage in the top 5 cm of the soil profile (Alvarez and Alvarez 2000). Dao (1998) indicated that cultivation, high temperature, and a semiarid climate accelerated organic carbon loss and weakened soil structure in the Southern Plains, and tillage and residue incorporation enhanced C mineralization and atmospheric fluxes, suggesting tillage intensity should be decreased to reduce C loss.

Tillage operations strongly control the soil environment by altering the soil geometry. These effects influence many physical, chemical and biological properties of the soil and thereby the conditions for crop growth. Alvarez and Alvarez (2000) stated that conservation tillage, especially no-tillage, induced changes in the distribution of organic pools in the soil profile. SOC gains under no-till were about 250 kg/ha/yr greater than for tilled systems regardless of cropping frequency in Canadian prairies with climate in semiarid conditions (Campbell *et al.,* 2005). Within the surface 7.5 cm, the no-till system possessed significantly more SOC (by 7.28 Mg/ha), particulate organic matter C (by 4.98 Mg/ha), potentially mineralizable N (by 32.4 kg/ha), and microbial biomass C (by 586 kg/ha), as well as greater aggregate stability (by 33.4%) and faster infiltration rates (by 55.6 cm/h) relative to the conventional tillage (Liebig *et al.,* 2004). Balota *et al.,* (2003) showed that no-tillage increased microbial biomass C, N, and P, and higher levels of more labile C existed in no-tillage systems than in conventional systems. However, conversion to zero tillage may not always result in an increase in soil C or N without adequate fertility (Campbell *et al.,* 2001a).

Much of the decomposition, as a result of soil tillage, takes place immediately after the soil is plowed and is directly related to the volume of soil disturbed and the roughness of the surface after plowing (Dao *et al.,* 2002). The moldboard plow causes the largest amount of carbon losses, while a deep tillage tool that does not invert soil causes little decomposition of organic matter. Any reduction of soil disturbance can be expected to reduce soil carbon losses. Conversely, the techniques of minimum tillage and no-till mean slower rates of organic matter decay (Acquaah 2002). In a University of Guelph experiment comparing different tillage practices for corn, the topsoil (0-15 cm depth) of plots devoted to no-till corn production had an average of 19.8 t/ha more organic matter after 18 years of research than did the plots where corn was grown using traditional tillage methods (fall moldboard plowing plus secondary tillage in spring time) (Vyn and Raimbault 1993). In this experiment, no difference was found among tillage treatments in soil organic matter levels at lower soil depths. More SOC was found in the top 10 cm and less in the 10-20 cm soil depth of the chisel plow than in the moldboard plow soils. However, chisel plow did not increase the SOC content (0-20 cm) above that of moldboard plow indicating this form of reduced tillage did not increase C sequestration in any of the rotations (Yang and Kay 2001).

After 11 years of different tillage operations in Chinese Mollisols, Liu *et al.*, (2005) reported that integrated tillage, where tillage system varied with each crop in the rotation (i.e., moldboard plow for wheat, deep chisel for corn and rotatory plow for soybean), had the highest levels of SOC and N in the upper soil layer in the Chinese Mollisol. Moldboard plowing had the lowest level of SOC and N content in the profile, with the largest reduction being in the top two layers. The SOC and N contents at 16-30 cm in the rotary plowing and conventional tillage were higher than in the depth between 0 and 15 cm, indicating that more root residues were incorporated into this layer. This result was consistent with mixing of organic matter by plowing but opposite to results with no-tillage practice or conservation tillage (Arshad *et al.*, 1990, Dalal *et al.*, 1991, Ding *et al.*, 2002). Overall, integrated tillage appeared effective in maintaining SOC and, maybe, soil productivity. Yang *et al.*, (2003a) indicated that the conversion from conventional tillage to conservation tillage at an annual rate of 2%, particularly no till, could reverse the loss of SOC in Chinese Mollisols within 20 years. However, this positive effect of conservation tillage on SOC in black soil area of China was only effective in severely eroded soil or in the farmland with slope, and it was not effective in flat and low-damp farmland.

It is thus evident that tillage practices have a major effect on soil properties, the distribution of C and N, and the rate of organic matter decomposition and N mineralization. The adoption of conservation tillage for reversing the decline of SOC in agricultural lands is possible in the black soil area of China, as it has been in many other countries. Continuous monitoring of long-term changes in the soil organic carbon and soil quality under conservation tillage in different agro-ecological zones is essential. There is also a need to obtain more data on long-term effects of different tillage systems on carbon and nitrogen mineralization and immobilization in field situations.

Crop Rotation

Crop rotation can have a major impact on soil health, due to emerging soil ecological interactions and processes that occur with time. These include improving soil structural stability and nutrient use efficiency, increasing crop water úse efficiency and soil organic matter levels, providing better weed control, and disrupting insect and disease life cycles (Carter *et al.*, 2002, Carter *et al.*, 2003). Crop rotations can also increase yields, and enhance nitrogen availability when nitrogen-fixing legumes are included (Galantini *et al.*, 2000, Miglierina *et al.*, 2000). Crop rotation systems are more effective effective at reducing long-term yield variability than monoculture systems, and can increase total soil and N concentrations over time, which may further improve soil productivity (Varvel 2000, Kelley *et al.*, 2003).

Carter *et al.*, (2003) indicated that losses of SOC during 11-year period ranged from marginal (4%) for rotations with Italian ryegrass, to significant 16%) under barley rotation, which illustrates the importance of C inputs in maintaining soil organic matter levels. Blair and Crocker (2000) examined the effect of different rotations, including legumes and fallows on soil structural stability, unsaturated hydraulic conductivity, and the concentration of different carbon fractions in a long-term rotation trial, and found that the inclusion of legume crops in the rotation resulted in an increase in liable carbon concentrations compared with continuous wheat or a long fallow period.

In comparison of maize-rice and rice-rice cropping systems, Witt *et al.*, (2000) found that the replacement of dry season rice by maize caused reduction in soil C and N due to a 33-41% increase in the estimated amount of mineralized and N during the dry season. As a result, there was 11-12% more C sequestration and 5-12% more N accumulation in soils continuously cropped with rice than in the maize-rice rotation with the greater amounts sequestered in N-fertilized treatments. Their results documented the capacity of continuous, irrigated rice systems to sequester C and N during relatively short time periods.

Yang and Kay (2001) found that continuous alfalfa had the greatest average SOC concentration 0-40 cm), and rotations had higher SOC concentration than continuous corn. Huggins *et al.,* (1998) reported that in the treatments containing both crops aboveground C returned to the soil from corn was in average 40% higher than C returned from soybean. Although more aboveground C was returned with corn, SOC did not differ with crop sequence or depth. Smith *et al.,* (2000) developed dynamic soil quality model to evaluate optimum cropping systems in the northern Great Plains, and indicated that a crop production system with continuous spring wheat and direct planting was the most profitable system and had lower soil erosion and higher soil quality attributes.

During a 9-year crop rotation experiment in the Chinese Mollisol, Liu *et al.,* (2003) found that cropping systems significantly altered SOC concentration. The SOC in the treatments of the wheat-sweet clover and wheat-soybean with addition of pig manure or wheat straw was significantly higher than that of the commonly used wheat-soybean rotation (wheat straw removed), particularly in the 0-17 cm horizon. For the overall SOC concentration (means of SOC of all three horizons), soil with addition of wheat straw was 22% greater than that of wheat-soybean alone, and similar differences occurred for overall SOC in the wheat-sweet clover rotation and wheat-soybean rotation with addition of pig manure. However, all three treatments increased SOC content in the 18-32 cm soil depth relative to the wheat-soybean rotation.

Liu *et al.,* (2003) also showed that the wheatsweet clover rotation not only increased the SOC content in all soil depths, but also had the greatest amount of SOC and had a decrease in soil bulk density. The wheat-soybean rotation with addition of wheat straw had a total of SOC (62 300 kg/ha), or an increase of 7.5% as compared with SOC in the typical wheat-soybean rotation in the region. The total SOC storage (sum of all three horizons) increase was 10.7% for wheat-soybean rotation with addition of manure, and 14.4% for wheatsoybean rotation with addition of wheat straw. The total amount of SOC increase (11 700 kg/ha) in wheat-soybean rotation with addition of wheat straw would correspond to sequestration of approximately 43 tons CO2/ha from atmosphere. Fang *et al.,* (2005) indicated that 2.24-5.0 Tg C was released into atmosphere annually ever since the cultivation of black soil in China, and maximum soil carbon sequestration potential could be as much as 1.55 Tg if appropriate measures were taken especially in crop rotations.

These results indicated that improved crop rotation strategies can increase the organic carbon reserve and improve soil structure and quality of the black soils, thereby sequestrating CO2 from the atmosphere and thus mitigating against the greenhouse effect. Further, the adoption of appropriate crop rotations to increase the quantity and quality of soil organic matter and hence soil chemical and physical will help to ensure the longterm sustainability of agriculture in the world.

Monoculture

The impacts of continuous monoculture systems on SOC, soil function and sustainability differed among crops (Russell and Jones 1996, Ryszkowski *et al.*, 1998). Acosta-Martinez *et al.*, (2004) concluded that continuous monoculture systems had a negative impact on soil function and sustainability. They found that organic C was higher in perennial pasture compared with continuous cotton at the depth of 0-5 cm, and soil microbial biomass, C, N and soil enzymes were also lower in the continuous cotton. In a Brown Chernozem, continuous wheat increased soil organic N in the 0-15 cm soil layer by 7-17 kg N/ha/yr more than fallow/wheat, and was accompanied by an increase in the proportion of mineralizable soil organic N (Liang *et al.*, 2004). Potter *et al.*, (1997) showed that total SOC content in the surface 20 cm increased by 5.6 t C/ha in the continuous wheat no-till treatment compared with the stubble mulch treatment. Campbell *et al.*, 2005) reported that replacing wheat with lentil had little effect on SOC gains, and replacing wheat with lower-yielding flax reduced SOC gains, while replacing wheat with erosion-preventing fall rye increased SOC gains. A large increase in SOC level for a rotation compared to a monoculture maize system was found by Gregorich *et al.*, (2001). Gil and Fick (2001) indicated that soil inorganic N of alfalfa monoculture was three- to five-fold higher than the alfalfa, red clover, and gamagrass-alfalfa mixture, but 30-50% lower than in gamagrass monoculture. Havlin *et al.*, (1990) assessed SOC dynamics in: (1) continuous sorghum (*Sorghum bicolor* L.), continuous soybean (*Glycine max* L.), and sorghum-soybean rotations combined with tillage and N fertilization; and (2) continuous corn with corn-soybean rotation. The high residue-producing continuous sorghum in the first and the continuous corn in the second rotation combined with reduced tillage and surface residue maintenance resulted in more SOC sequestration than grain-legume rotations. Similar results were obtained by Omay *et al.,* (1997) who reported more SOC under continuous corn than under corn-soybean rotation. The results by Yang *et al.,* (2003b) in Chinese Mollisols also supported this conclusion. Drury *et al.,* (1998) compared continuous corn with continuous Kentucky bluegrass (*Poa pratensis* L.) and 4-year corn-oat (*Avena sativa* L.)-alfalfa-alfalfa (*Medicago sativa* L.) rotation. The SOC level was in the order bluegrass > 4-year rotation > continuous corn.

Gajda and Martyniuk (2005) showed that the activities of the dehydrogenase and phosphatase and microbial biomass C and N contents in the monoculture soil were significantly lower than those in the soil from the organic and conventional-short rotation systems. Ryszkowski *et al.,* (1998) reported that continuous cultivation of rye led to fauna impoverishment, and increased the number of crop pests and fungi with bacteria, and that actinomycetes became less abundant. Besides, the average concentration of phenolic acids in the soil of continuous rye was 400% higher than under rye in a diversified rotation. However, Gregorich *et al.,* (2001) reported that chemical composition of organic matter was little affected by the nature of crop residues.

Liu *et al.,* (2005) reported that for the Chinese Mollisols, continuous cropping of wheat, corn and soybean, compared to a rotation, reduced soil C and N contents at all depths in the profile with the exception of N content in continuous wheat. These results were similar to those of Mikhailova *et al.,* (2000). Continuous wheat had the least impact on soil C and N contents. The total microbial C of topsoil layer was 533 mg/kg in continuous wheat, 429 mg/kg in the crop rotation, but only 350 mg/kg for continuous corn and 398 mg/kg for continuous soybean. Thus, continuous corn and soybean not only reduced SOC and N but also led to a decline of soil biological activity. Collins *et al.,* (1992) reported a greater amount of soil organic matter and soil microbial biomass in continuous wheat than in a wheat-fallow rotation after 58 years. Liu *et al.,* (2005) indicated that 11-year continuous corn, soybean, and wheat resulted in a 10%, 11% and 5.3% decline in SOC, respectively, compared with a rotation of all crops; continuous cropping also resulted in a 6.5% decline in N for corn and soybean but there was no decline for wheat. These results suggest that if continuous corn, soybean, or wheat were to be adopted on Chinese Mollisols, an average annual decline rate of soil carbon in the 0-90 cm soil profile would be 0.91%, 0.97% and 0.48%, respectively. The decline in soil N would be 0.53% for continuous soybean or corn. Another research showed that a decline in SOC significantly reduced the N supply and resulted in the deterioration of soil physical properties (Stevenson, 1994).

Liu (2004) reported that for Chinese Mollisols continuous soybean had a greater decline in SOC and N contents at the 51-70 cm and 71-90 cm depths than the other treatments. This effect may be due to the tap root system of soybean as well as harmful impacts of continuous cropping on soybean root nodules and nitrogen fixation (Liu and Herbert, 2002). Liu (2004) also found declines in soil pH and of phosphatase, urease, invertase and dehydrogenase activities, but an increase in sucrase activity in continuous soybean soil. Moreover, the number of bacteria in continuous soybean was significantly lower than in normal soybean rotations, whereas fungi were more numerous, resulting in a decline in the bacteria/fungi ratio. This imbalance in the microbe populations in soybean monoculture may contribute to the decline in soil fertility. A higher bacteria/fungi ratio is consistent with higher soil fertility, whereas a lower ratio favoring fungi is consistent with a low fertility soil. Liu and Herbert (2002) proposed that some acid compounds were deposited in the continuous soybean soil, contributing to the deterioration of soil biological activity. Hence, the increase and influence of fungi in the continuous soybean soil rhizosphere demonstrates that continuous soybean is not sustainable unless countermeasures are taken to deal with the fungi increase.

Crop type, therefore, influences SOC and soil function in continuous monoculture systems. The negative prominent impacts of monoculture are fauna impoverishment, increased number of crop pests, declined activities of dehydrogenase and phosphatase, and higher phenolic acids in the soil. Although continuous wheat increased microbial biomass and alfalfa haycrop increased inorganic N significantly, continuous monoculture is not sustainable for many crops unless countermeasures are taken to deal with.

Fertilization

Fertilization is one of the most important practices in crop production for its influence on soil nutrients availability. Ishaq *et al.*, (2002) showed that a fertilizer application significantly increased soil P and K concentrations, and the concentrations of N, P, K and SOC were greater in the plough layer than in the subsoil. Nitrogen is the nutrient most limiting to crop production in all areas of the world and is generally applied to soil in a large quantity. Since the application of N to the soil is subject to losses by volatilization, immobilization, denitrification and leaching, it may be necessary to compensate this by adjusting the fertilizer management. Fertilizer use efficiency may also change with changes in tillage management. Malhi *et al.*, (2001) indicated that placing the fertilizer in a band reduced contact with soil microorganisms, and reducing immobilization of both ammonium (NH_4 +) and nitrate (NO_3 –). Banding also slowed down the conversion of urea to NH_3 and NH_4 + to NO_3 –, which can reduce N losses by volatilization and leaching. Reducing tillage intensity modified both the demand of crops for N due to changes in yield potential, and supply of N due to changes in N cycling and losses. The N fertilization effects on SOC were most evident when stover was returned to no-till plots (Clapp *et al.*, 2000). Farmyard manure and the recycling of crop residues with NPK supplementation are efficient ways of fertilizing maize and wheat. Significantly higher yields were obtained at high levels of NPK fertilization, especially in rotations where the proportion of maize or wheat was 50% or higher (Berzsenyi *et al.*, 2000). Hao *et al.*, (2002) showed that the effects of manure application, tillage, crop rotation, fertilizer rate, and soil and water conservation farming on SOC pool were cumulative.

No tillage continuous corn with NPK, lime, and cattle manure was an effective cropland management system for SOC sequestration. Campbell *et al.*, (2000) found that SOC was increased most by annual cropping with application of adequate fertilizer N and P in semiarid southwestern Saskatchewan. They also found that soil organic C and total N, microbial biomass, light fraction organic C and N, mineralizable N and wet aggregate stability, generally had positive responses to fertilization (Campbell *et al.*, 2001b). Reddy *et al.*, (2003) showed that continuous application of NPK plus farmyard manure led to a marked increase in organic C and total N as compared to an adjacent areas of left fallow in an Eutrochrept soil. The introduction of crop residue mulch and higher rates of fertilizers were ecommended for sustaining soil quality (Sarno *et al.,* 2004). However, the effects of fertilization on soil C were small, and differences were observed only in the monoculture maize system (Gregorich *et al.,* 2001).

After 16 years of different fertilization treatments in the crop rotation, Liu *et al.,* (2005) found that the profile average SOC content (0-90 cm) was only 0.9%, 4.1%, and 8.6% higher for manure, chemical fertilizers, and manure plus fertilizers, respectively, than that with no fertilizer application or control in the Chinese Mollisols. However, SOC at the 0-15 cm soil layer was 6.2%, 7.7%, and 9.3% higher with manure, chemical fertilizers, and manure plus fertilizers, respectively, than with no fertilizer application. These results indicated that the annual rate of decline rate of SOC in the 0-15 cm layer without fertilizer was not very high (<0.58%/yr) when a well-designed crop rotation was used. The results were similar to data from long-term experiments in Denmark and England that revealed a slow change in SOC levels under temperate conditions in response to changes in different land uses (Christensen and Johnson 1997). Yang *et al.,* (2003b) further indicated that the SOC content could be maintained at a relatively stable level under sufficient chemical fertilizer application application without return of manure and crop residue conditions, and SOC content was increased at the combination of chemical fertilizer and manure application. This indicates that corn residue and exudates themselves could keep SOC equilibrium under current production level and management practices.

Liu *et al.,* (2005) also reported for the Chinese Mollisols that manure alone did not increase the N content in the soil profile compared to that of no fertilizer application in the crop rotation. However, chemical fertilizers and manure plus fertilizers significantly increased N contents, especially at the 0-15 cm and 16-30 cm soil layers. Francioso *et al.,* (2000) also reported that SOC and N differed significantly after 22 years for all treatments where the amendments with cattle manure markedly increased the SOC and N contents, while cow slurries and crop residues reduced SOC and N contents. The greatest reduction for SOC and N contents was in the unamended plots after 22 years. Reeves (1997) suggested that soil organic matter can be preserved only by ley rotations with reduced tillage frequency. Generally, an adequate application of fertilizers combined with farmyard manure could increase soil nutrients, and SOC content in the Chinese Mollisols. Manure or crop residue alone could not maintain SOC levels, and SOC can only be preserved by rotation with reduced tillage frequency and additions of chemical fertilizers and manure.

Summary

The dynamic processes that influence soil quality are complex, and they operate through time at different locations and situations. Soil organic matter is both a source of carbon release and a sink for carbon sequestration. Cultivation and tillage can reduce and change the distribution of SOC while an appropriate crop rotation can increase or maintain the quantity and quality of soil organic matter, and improve soil chemical and physical properties. The return of crop residues and the application of manure and fertilizers can all contribute to an increase in soil nutrients and SOC content, but would need to be combined into a management system for more improvement. The negative prominent impacts of monoculture are influenced by crop type with fauna impoverishment, an increased number of crop pests, a decline in activities of dehydrogenase and phosphatase, and increased levels of phenolic acids in the soil. SOC can only be preserved by using crop rotations with reduced tillage frequency and additions of chemical fertilizers, crop residues and/or manure. Continuous monitoring of long-term changes in the SOC and soil quality under conservation tillage in different agro-ecological zones is essential. There is also a need to obtain more data on longterm effects of different tillage systems on carbon and nitrogen mineralization and immobilization in various field situations. The issue involved in understanding soil quality and the design of crop and soil systems for agricultural sustainability should be more holistic, and it needs further investigation.

(Xiaobing Liu, Key Laboratory of Black Soil Ecology, Northeast Institute of Geography and Agroecology, Chinese Academy of Sciences, Harbin, PR China. The author can be reached at xbxliu@yahoo.com,

S J Herbert, Department of Plant, Soil and Insect Sciences, University of Massachusetts, Amherst, USA,

A M Hashemi, Department of Plant, Soil and Insect Sciences, University of Massachusetts, Amherst, USA,

X Zhang, Key Laboratory of Black Soil Ecology, Northeast Institute of Geography and Agroecology, Chinese Academy of Sciences, Harbin, PR China,

G Ding, Chemistry Department, Northern State University, Aberdeen, USA.)

References

Acosta-Martinez V., Zobeck T.M., Allen V. (2004): "Soil microbial, chemical and physical properties in continuous cotton and integrated crop-livestock systems". *Soil Sci. Soc. Am. J.*, 68: 1875-1884.

Acquaah G. (2002): *Principles of Crop Production: Theory, Techniques, and Technology*. Pearson Education, Inc., Upper Saddle River, New Jersey.

Alvarez C.R., Alvarez R. (2000): "Short-term effects of tillage systems on active soil microbial biomass". *Biol. Fert. Soils*, 31: 157-161.

Arshad M.A., Coen G.M. (1992): "Characterization of soil quality: Physical and chemical criteria". *Am. J. Alter. Agr.*, 7: 25-30.

Arshad M.A., Martin S. (2002): "Identifying critical limits for soil quality indicators in agro-ecosystems". *Agr. Ecosyst. Environ.*, 88: 153–160.

Arshad M.A., Schnitzer M., Angers D.A., Ripmeester J.A. (1990): "Effects of till vs no-till on the quality of soil organic matter". *Soil Biol. Biochem.*, 22: 595-599.

Bakker H.J.I. (ed.) (1990): *The World Food Crisis: Food Security in Comparative Perspective*. Can. Scholars' Press, Toronto.

Balota E.L., Colozzi A., Andrade D.S., Dick R.P. (2003): "Microbial biomass in soils under different tillage and crop rotation systems". *Biol. Fert. Soils*, 38: 15-20.

Bauer A., Black A.L. (1994): "Quantification of the effect of soil organic matter content on soil productivity". *Soil Sci. Soc. Am. J.*, 58: 185-193.

Beare M.H., Cabrera M.L., Hendrix P.F., Coleman D.C. (1994): "Aggregate-protected and unprotected organic matter pools in conventional and no-tillage soils". *Soil Sci. Soc. Am. J.*, 58: 787-795.

Berzsenyi Z., Gyorffy B., Lap D. (2000): "Effect of crop rotation and fertilization on maize and wheat yields and yield stability in a long-term experiment". *Eur. J. Agron.*, 13: 225-244.

Blair N., Crocker G.J. (2000): "Crop rotation effects on soil carbon and physical fertility of two Australian soils". *Aust. J. Soil Res.*, 38: 71-84.

Brady N.C., Weil R.R. (2000): *Elements of the Nature and Properties of Soils*. Prentice-Hall, Inc., Upper Saddle River, New Jersey.

Brady N.C., Weil R.R. (2002): *Nature and Properties of Soils*. Prentice-Hall, Inc., Upper Saddle River, New Jersey.

Campbell C.A. (1998): "Possibilities for future carbon sequestration in Canadian agriculture in relation to land use changes". *Clim. Changes,* 40: 81-103.

Campbell C.A., Janzen H.H., Paustian K., Greegorich E.G., Sherrod L., Liang B.C., Zentner R.P. (2005): "Carbon storage in soils of the North American Great Plains: Effect of cropping frequency". *Agron. J.,* 97: 349-363.

Campbell C.A., Selles F., Lafond G.P., Biederbeck V.O., Zentner R.P. (2001a): "Tillage-fertilizer changes: Effect on some soil quality attributes under long-term crop rotation in a thin Black Chernozem". *Can. J. Soil Sci.,* 81: 157–165.

Campbell C.A., Selles F., Lafond G.P., Zentner R.P. (2001b): "Adopting zero tillage management: Impact on soil C and N under long-term crop rotations in a thin Black Chernozem". *Can. J. Soil Sci.,* 81: 139-148.

Campbell C.A., Zentner R.P., Liang B.C., Roloff G., Gregorich E.C., Blomer B. (2000): "Organic C accumulation in soil over 30 years in semiarid southwestern Saskatchewan – Effect of crop rotations and fertilizers". *Can. J. Soil Sci.,* 80: 179-192.

Carter M.R., Kunelius H.T., Sanderson J.B., Kimpinski J., Platt H.W., Bolinder M.A. (2003): "Productivity parameters and soil health dynamics under long-term 2-year potato rotation in Atlantic Canada". *Soil Till. Res.,* 72: 153-168.

Carter M.R., Sanderson J.B., Ivany J.A., White R.P. (2002): "Influence of rotation and tillage on forage maize productivity, weed species, and soil quality of a fine sandy loam in the cool-humid climate of Atlantic Canada". *Soil Till. Res.,* 67: 85-98.

Chan K.Y., Heenan D.P., So H.B. (2003): "Sequestration of carbon and changes in soil quality under conservation tillage on light-textured soils in Australia: A review". *Aust. J. Exp. Agr.,* 43: 325-334.

Christensen B., Johnson A.E. (1997): "Soil organic matter and soil quality-lessons learned from long-term experiments at Askov and Rothamsted". In: Gregorich E.G., Carter M.R. (eds.): *Soil Quality for Crop Production and Ecosystem Health.* Developments in Soil Science 25. Elsevier, Amsterdam: 399-430.

Clapp C.E., Allmaras R.R., Layese M.F., Linden D.R., Dowdy R.H. (2000): "Soil organic carbon and 13C abundance as related to tillage, crop residue, and nitrogen fertilization under continuous corn management in Minnesota". *Soil Till. Res.,* 55: 127-142.

Collins H.P., Rausmussen P.E., Douglas C.L. Jr. (1992): "Crop rotation and residue management effects on soil carbon and microbial dynamics". *Soil Sci. Soc. Am. J.,* 56: 783-788.

Dalal R.C., Henderson P.A., Glasby J.M. (1991): "Organic matter and microbial biomass in a Vertisol after 20 yr. of zero-tillage". *Soil Biol. Biochem.*, 23: 435-441.

Dao T.H. (1998): "Tillage and crop residue effects on carbon dioxide evolution and carbon storage in a Paleustoll". *Soil Sci. Soc. Am. J.*, 62: 250-256.

Dao T.H., Stiegler J.H., Banks J.C., Boerngen L.B., Adams B. (2002): "Post-contrast use effects on soil carbon and nitrogen in conservation reserve grasslands". *Agron. J.*, 94: 146-152.

Davidson E.A., Ackerman I.L. (2003): "Changes in soil carbon inventories following cultivation of previously untilled soil". *Biogeochemistry*, 20: 161-193.

Ding G., Novak J.M., Amarasiriwardena D., Hunt P.G., Xing B. (2002): "Soil organic matter characteristics as affected by tillage management". *Soil Sci. Soc. Am. J.*, 66: 421-429.

Doran J.W., Parkin T.B. (1994): "Defining and assessing soil quality". In: Doran J.W. *et al.*, (eds.): *Defining Soil Quality for a Sustainable Environment.* SSSA Spec. Publ. No. 35, ASA and SSSA, Madison, WI: 3-21.

Doran J.W., Smith M.S. (1987): "Organic matter management and utilization of soil and fertilizer nutrients". In: Follett R.F. *et al.*, (eds.): *Soil Fertility and Organic Matter as Critical Components of Agricultural Production Systems.* SSSA Spec. Publ. No. 19, ASA and SSSA, Madison, WI: 53-72.

Drury C.F., Oloya T.O., McKenney D.J., Gregorich E.G., Tan C.S., van Luyk C.L. (1998): "Long-term effects of fertilization and rotation on denitrification and soil carbon". *Soil Sci. Soc. Am. J.*, 62: 1572-1579.

Dumanski J., Desjardins R.L., Tarnocai C., Monreal D., Gregorich E.G., Kirkwood V., Campbell C.A. (1998): "Possibilities for future carbon sequestration in Canadian agriculture in relation to land use changes". *Clim. Changes*, 40: 81-103.

Fang H.J., Yang X.M., Zhang X.P., Liang A.Z. (2005): "Using 137 Cs tracer technique to evaluate soil erosion and deposition of black soil in northeast China". *J. Appl. Ecol.*, 16: 464-468 (In Chinese).

Farquharson R.J., Schwenke G.D., Mullen J.D. (2003): "Should we manage soil organic carbon in Vertosols in the northern grains region of Australia?" *Aust. J. Exp. Agr.*, 43: 261-270.

Feichtinger F., Erhart E., Hartl W. (2004): "Net N-mineralisation related to soil organic matter pools". *Plant Soil Environ.*, 50: 273-276.

Follett (2001): "Soil management concepts and carbon sequestration in cropland soils". *Soil* Till. *Res.*, 61: 77-92.

Francioso O., Ciavatta C., Sanche-Cortes S., Tugnoli V., Sitti L., Gessa C. (2000): "Spectroscopic characterization of soil organic matter in long-term amendments trials". *Soil*

Sci., 165: 495-504.

Freixo A.A., Machado Plod, dos Santos H.P., Silva C.A., Fadigas F.D. (2002): "Soil organic carbon and fractions of a Rhodic Ferralsol under the influence of tillage and crop rotation systems in southern Brazil". *Soil Till. Res.*, 64: 221-230.

Gajda A., Martyniuk S. (2005): "Microbial biomass C and N and activity of enzymes in soil under winter wheat grown in different crop management systems". *Pol. J. Environ. Stud.*, 14: 159-163.

Galantini J.A., Landriscini M.R., Iglesias J.O., Miglierina A.M., Rosell R.A. (2000): "The effects of crop rotation and fertilization on wheat productivity in the Pampean semiarid region of Argentina. II. Nutrient balance, yield and grain quality". *Soil Till. Res.*, 53: 137-144.

Gardiner D.T., Miller R.W. (2004): *Soils in Our Environment.* 10th ed. Prentice-Hall, Inc., Upper Saddle River, New Jersey.

Gil J.L., Fick W.H. (2001): "Soil nitrogen mineralization in mixtures of eastern gamagrass with alfalfa". *Agron J.*, 93: 902-910.

Gregorich E.G., Drury C.F., Baldock J.A. (2001): "Changes in soil carbon under long-term maize in monoculture and legume-based rotation". *Can. J. Soil Sci.*, 81: 21-31.

Hammond A.L. (ed.) (1992): *World Resources 1992-93.* Oxford Univ. Press, Oxford.

Hao Y., Lal R., Owens L.B., Lzaurralde R.C., Post W.M., Hothem D.L. (2002): "Effect of cropland management and slope position on soil organic carbon pool at the North Appalachian Experimental Watersheds". *Soil Till. Res.*, 68: 133-142.

Havlin J.L., Kissel D.E., Maddux L.D., Claassen M.M., Long J.H. (1990): "Crop rotation and tillage effects on soil carbon and nitrogen". *Soil Sci. Soc. Am. J.*, 54: 448-452.

Hillel D.J. (1991): "Out of the earth-civilization and the life of the soil". The Free Press, New York, NY.

Huggins D.R., Clap C.E., Allmaras R.R., Lamb J.A., Layese M.F. (1998): "Carbon dynamics in corn-soybean sequences as estimated from natural carbon-13 abundance". *Soil Sci. Soc. Am. J.*, 62: 195-203.

Ishaq M., Ibrahim M., Lal R. (2002): "Tillage effects on soil properties at different levels of fertilizer application in Punjab, Pakistan". *Soil Till. Res.*, 68: 93-99.

Jastrow J.D., Boutton T.W., Miller R.M. (1996): "Carbon dynamics of aggregate-associated organic matter estimated by carbon-13 natural abundance". *Soil Sci. Soc. Am. J.*, 60: 801-807.

Jenny H. (1980): *The Soil Resource: Origin and Behavior.* Ecol. Stud. 37. Springer-Verlag, New York.

Kelley K.W., Long J.H., Todd T.C. (2003): "Long-term crop rotations affect soybean yield, seed weight, and soil chemical properties". *Field Crops Res.*, 83: 41-50.

Kern J.S., Johnson M.G. (1993): "Conservation tillage impacts on national soil and atmospheric carbon levels". *Soil Sci. Soc. Am. J.*, 57: 200-210.

Kladivko E.J. (2001): "Tillage systems and soil ecology". *Soil Till. Res.*, 61: 61-76.

Kushwaha C.P., Tripathi S.K., Singh K.P. (2001): "Soil organic matter and water-stable aggregates under different tillage and residue conditions in a tropical dryland agroecosystem". *Appl. Soil Ecol.*, 16: 229-241.

Lal R., Kimble J., Follett R. (1997): "Soil quality management for carbon sequestration". In: Lal R. *et al.*, (eds.): *Soil Properties and their Management for Carbon Sequestration.* US Dep. Agr., Nat. Res. Conserv. Serv., Nat. Soil Surv. Cent., Lincoln, NE: 1-8.

Lal R., Kimble J., Levine E., Whitman C. (1995): "World soils and greenhouse effect: An overview". In: Lal R. *et al.*, (eds.): *Soils and Global Change.* Lewis Publ., Boca Raton, FL: 1-8.

Larson W.E., Pierce F.J. (1991): "Conservation and enhancement of soil quality". In: Mumanski J. *et al.*, (eds.): *Evaluation for Sustainable Land Management in the Developing World.* Vol. 2. Techn. Pap. In: Proc. Int. Workshop, Chiang Rai, Thailand, Int. Board Soil Res. Manage. Bangkok: 175-203.

Liang B.C., McConkey B.G., Campbell C.A., Curtin D., Lafond G.P., Brandt S.A., Moulin A.P. (2004): "Total and labile soil organic nitrogen as influenced by crop rotations and tillage in Canadian prairie soils". *Biol. Fert. Soils*, 39: 249-257.

Liebig M.A., Tanaka D.L., Wienhold B.J. (2004): "Tillage and cropping effects on soil quality indicators in the northern Great Plains". *Soil Till. Res.*, 78: 131-141.

Liu X.B. (2004): "Changes in soil quality under different agricultural management in the Chinese Mollisols". [Ph.D. Dissertation.] Univ. Massachusetts, Amherst, MA.

Liu X.B., Han X.Z., Herbert S.J., Xing B. (2003): "Dynamics of soil organic carbon under different agricultural management systems in the black soil of China". *Commun. Soil Sci. Plant Anal.*, 34: 973-984.

Liu X.B., Herbert S.J. (2002): "Fifteen years of research examining cultivation of continuous soybean in Northeast China". *Field Crops Res.*, 79: 1-7.

Liu X.B., Liu J.D., Xing B., Herbert S.J., Zhang X.Y. (2005): "Effects of long-term continuous cropping, tillage, and fertilization on soil carbon and nitrogen in Chinese Mollisols". *Commun. Soil Sci. Plant Anal.*, 36: 1229-1239.

Loveland P., Webb J. (2003): "Is there a critical level of organic matter in the agricultural soils of temperate regions: A review". *Soil Till. Res.*, 70: 1-18.

Magdoff F., van Es H. (2000): *Building Soils for Better Crops*. Univ. Vermont, Burlington VT, USA: 9-13.

Malhi S.S., Grant C.A., Johnston A.M., Gill K.S. (2001): "Nitrogen fertilization management for no-till cereal production in the Canadian Great Plains: A review". *Soil Till. Res.*, 60: 101-122.

Mausbach M., Tugel A. (1995): "Decision document for establishing a Soil Quality Institute". White Pap. Nat. Res. Conserv. Serv.

Miglierina A.M., Iglesias J.O., Landriscini M.R., Galantini J.A., Rosell R.A. (2000): "The effects of crop rotation and fertilization on wheat productivity in the Pampean semiarid region of Argentina. I. Soil physical and chemical properties". *Soil Till. Res.*, 53: 129-135.

Mikhailova E.A., Bryant R.B., Vassenev I.I., Schwager S.J., Post C.J. (2000): "Cultivation effects on soil carbon and nitrogen contents at depth in the Russian Chernozem". *Soil Sci. Soc. Am. J.*, 64: 738-745.

Omay A.B., Rice C.W., Maddux L.D., Gordon W.B. (1997): "Changes in soil microbial and chemical properties under long-term crop rotation and fertilization". *Soil Sci. Soc. Am. J.*, 61: 1672-1678.

Onemli F. (2004): "The effects of soil organic matter on seedling emergence in sunflower (*Helianthus annuus* L.)". *Plant Soil Environ.*, 50: 494-499.

Paustian K., Andren O., Janzen H.H., Lal R., Smith P., Tian G., Tiesen H., Van Smith M.P., Powlson D.S., Glendining M.J., Smith J.U. (1998): "Preliminary estimates of the potential for carbon mitigation in European soils through no-till farming". *Glob. Change Biol.*, 4: 679-685.

Post W.M., Kwon K.C. (2000): "Soil carbon sequestration and land-use changes: Processes and potential". *Glob. Change Biol.*, 6: 317-327.

Potter K.N., Jones O.R., Torbert H.A., Unger P.W. (1997): "Crop rotation and tillage effects on organic carbon sequestration in the semiarid southern Great Plains". *Soil Sci.*, 162: 140-147.

Power J.F., Meyers R.J.K. (1989): "The maintenance or improvement of farming systems in North America and Australia". In: Stewart J.W.B. (ed.): *Soil quality in semi-arid agriculture*. In: Proc. Int. Conf. Univ. Saskatchewan, Saskatoon, Canada: 273-292.

Reddy K.S., Singh M., Tripathi A.K., Saha M.N. (2003): "Changes in amount of organic and inorganic fractions of nitrogen in an Eutrochrept soil after longterm cropping with different fertilizer and organic manure inputs". *J. Plant Nutr. Soil Sci., Z. Pfl.-Ernähr.* Bodenkde, 166: 232-238.

Reeves D.W. (1997): "The role of soil organic matter in maintaining soil quality in continuous cropping systems". *Soil Till. Res.*, 43: 131-167.

Robinson C.A., Cruse R.M., Kohler K.A. (1994): "Soil management". In: Hatfield J.L., Karlen D.L. (eds.): *Sustainable agricultural systems*. Lewis Publ., Boca Raton, FL: 109-134.

Roscoe R., Burman P. (2003): "Tillage effects on soil organic matter in the density fractions of a Cerrado Oxisol". *Soil Till. Res.*, 70: 107-119.

Russell J.S., Jones P.N. (1996): "Continuous, alternative and double crop systems on a Vertisol in subtropical Australia". *Aust. J. Exp. Agr.*, 36: 823-830.

Ryszkowski L., Szajdak L., Karg J. (1998): "Effects of continuous cropping of rye on soil biota and biochemistry". *Crit. Rev. Plant Sci.*, 17: 225-244.

Sarno M.Iijima, Lumbanraja J., Sunyoto, Yuliadi E., Izumi Y., Watanabe A. (2004): "Soil chemical properties of an Indonesian red soil as affected by land use and crop management". *Soil Till. Res.*, 76: 115-124.

Singer M.J., Ewing S.A. (2000): Soil quality. In: Sumner M.E. (ed.): *Handbook of Soil Science*. CRC Press, Inc., Boca Raton, FL: 271-298.

Smith E.G., Lerohl M., Messele T., Janzen H.H. (2000): "Soil quality attribute time paths: Optimal levels and values". *J. Agr. Res. Econ.*, 25: 307-324.

Sojka R.E., Upchurch D.R. (1999): "Reservations regarding the soil quality concept". *Soil Sci. Soc. Am. J.*, 63: 1039-1054.

Stevenson F.J. (1994): *Humus Chemistry: Genesis, Composition, Reactions.* (2nd ed.) John Wiley and Sons, New York.

Unger P.W. (1991): "Organic matter, nutrient, and pH distribution in no- and conventional – tillage semiarid soils". *Agron. J.*, 83: 186-189.

Varvel G.E. (2000): "Crop rotation and nitrogen effects on normalized grain yields in a long-term study". *Agron. J.*, 92: 938-941.

Vyn T.J., Raimbault B.A. (1993): "Long-term effects of 5 tillage systems on corn response and soil structure". *Agron. J.*, 85: 1074-1079.

West T.O., Post W.M. (2002): "Soil organic carbon sequestration rates by tillage and crop rotation: A global data analysis". *Soil Sci. Soc. Am. J.*, 66: 1930-1946.

Witt C., Cassman K.G., Olk D.C., Biker U., Liboon S.P., Samson M.I., Ottow J.C.G. (2000): "Crop rotation and residue management effects on carbon sequestration, nitrogen cycling and productivity of irrigated rice system". *Plant Soil*, 225: 263-278.

Yang X.M., Kay B.D. (2001): "Rotation and tillage effects on soil organic carbon sequestration in a typic Hapludalf in Southern Ontario". *Soil Till. Res.*, 59: 107-114.

Yang X.M., Zhang X.P., Deng W., Fang H.J. (2003a): "Black soil degradation by rainfall erosion in Jilin, China". *Land Degrad. Dev.*, 14: 409-420.

Yang X.M., Zhang X.P., Fang H.J., Zhu P., Ren J., Wang L.C. (2003b): "Effects of fertilization under continuous corn on organic carbon in black soil: Simulation by RothC-26.3 model". *Agr. Sci. China*, 36: 1318-1324 (In Chinese).

Zibilske L.M., Bradford J.M., Smart J.R. (2002): "Conservation tillage induced changes in

4

Agricultural Transformation
Towards Marketization

Nirbachita Karmakar

The article attempts to explore the marketization aspect of the process of agricultural transformation. First, it defines the process of agricultural transformation as a subset of structural transformation with individual farms going for a specialized production from a diversified subsistence production. Subsequently, it focuses on different features of agricultural transformation like structure of the agrarian systems, describing the process of agricultural transformation and specialization, marketing of the agricultural produce and marketing information system. Then it brings out the initiatives that can be taken to address the agricultural transformation towards marketization in the context of developing countries. Finally, it discusses one of the most important aspects of marketization, i.e., marketing information system with an illustration in the context of Indian agriculture.

Introduction

Conventionally, agricultural sector has been playing more of a supportive role in the process of economic development. Influenced by western counterparts, economic development is seen as a process of rapid structural transformation from a predominantly agricultural based economy to a complex industrial and service based society in the context of developing countries. In such a backdrop, agricultural sector supplemented the so-called dynamic industrial sector with sufficient supply of low-priced food and expanding labor power. Of late, the development economists have come to realize that apart from providing a supplementary role, agrarian sector and the rural economy can be indispensable in the overall strategy of the developing nations.

Now an aspect of structural transformation is agricultural transformation, defined as the process by which individual farms go for a specialized production from a diversified subsistence production with the objective of creating a more market-oriented agrarian economy. An agriculture and employment based strategy should encompass at least three basic components – a high rate of agricultural production through technological, institutional and price incentive mechanisms and by raising the productivity of small farmers; following an employment based urban development strategy can raise the domestic demand for agricultural output; lastly, the farming community can be supported through a diversified non-agricultural and labor-intensive rural activities. Thus agricultural transformation can be market-oriented through an integrated rural development. The former can be optimized by addressing certain issues:

Firstly, what should be the mechanism in boosting the agricultural output and per capita productivity that will benefit both the average small farmer as well as the landless rural labor and at the same time facilitating a sufficient food surplus for the emerging urban industrial sector?

Secondly, the process to transform low productivity farms to high end commercial enterprises needs to be explored.

Thirdly, decision-making process of the farmers, whether guided rationally or irrationally?

Fourthly, risk mitigation process of the small scale farmers of the low-income countries and the consequence of such risks faced by them.

Next, whether economic and price incentive mechanisms are sufficient to facilitate integrated rural development or does it call for institutional and structural change within the rural farming community.

Finally, how the rural development can be strengthened?

Structure of the Agrarian Systems

There are two distinct types of farming in the world agriculture. First, we have the well-developed agrarian sector of the developed nations with huge productive capacity and high output per worker. On the other hand, there is inefficient low productivity agriculture of the developing nations with agriculture hardly capable of sustaining the farm population even at a subsistence level. In between these two dominant set-ups of agrarian framework, there are some emerging economies of the developing countries like Punjab in India, some export-intensive Latin American and Asian countries achieving new milestone in agricultural production by undergoing agriculture transformation. However, the gap between the developed and the developing countries is all the more prevalent as evident in labor productivity. During the 60s, per capita agricultural output of the developed nations was about $680 while in it was only $52. In the new Millennium this productivity gap has further widened.

Agricultural Transformation and Specialization

Agricultural transformation in a way can signify a gamut of changes in the food production practices with intense usage of inputs as well as introduction of new farming practices and technological innovations.

Agricultural specialization, also known by the name of agricultural transformation, can be for particular product(s) like crop, fruits, livestock, cereals, pulses, milk, meat-cattle and organic honey, etc. There can also be functional specialization in agriculture, be it management, mechanization, plant protection, harvesting, guarding, risk management, and marketing. Again there can also be specialization of agricultural labor with coordination of specialized activity and exchange of products and services between the economic units like the household and the farm sectors. The coordination, specialization and exchange in agriculture sector can be facilitated by free market mechanisms like price movements, private negotiations or through a third party.

Agricultural transformation is specialization of different agricultural activities within markets or across markets. At one end, a high transaction costs could restrict specialization in small scale farming while a specialized farm entrepreneur carrying out the entire productive activity by purchasing all related agricultural services from specialized markets.

There can be different forms for managing the farming activity depending on the individual traits like skills, ability, effort and certain institutional, market and natural environment.

In this jet age it is almost imperative to think of transforming a traditional agrarian framework into a highly commercial farming system. In an attempt to commercialize agriculture and make it more market-oriented there had been introduction of cash crops resulting in peasant's loss of land to moneylenders or landlords. For small scale farmers, exclusive dependence on cash crops can be more detrimental than pure agriculture, the reason being attributed to price fluctuations due to uncertainty in nature.

Therefore, going for agriculture transformation, i.e., from subsistence to specialized, there is an intermediate step known as the diversified or mixed farming. This stage is dominated by cash crops like tea, coffee, specialized fruits and vegetables and animal husbandry with staple crop no longer dominating the farm output. The process of agricultural transformation can overcome the slack period of disguised unemployment. The situation is more observable in developing countries with abundant supply of cheap labor.

The outcome of above efforts to transform traditional agriculture to a commercial one is not only conditional on farmer's intrinsic ability and expertise in raising the agricultural productivity but also a social, commercial and an institutional set up. With the emergence of agricultural transformation the traditional farmer can easily respond to the new economic opportunities and incentives if he is assured of reasonable and reliable accessibility to credit, fertilizer, market information, fair market price for his farm output, and with reasonable benefits accruing to him and his family members.

After diversification the farm goes for specialization and thereby venturing into modern commercial farming. Agricultural transformation is acutely prevalent in industrial nations and evolving at par with structural transformation of the national economy. Improvement in the standard of living, biological and technological innovations, and expansion of local, national and international markets for a strong impetus on agricultural transformation. In agricultural transformation, when going for crop specialization, securing food for the family with some marketable surplus no longer remains the basic objective. Rather the basic goal rests purely on commercial profit with maximum per-hectare production; i.e., production is entirely market-oriented. Agricultural transformation emphasizes on capital formation, technological progress, scientific research and economic development.

Agricultural transformation varies in terms of size and function with regard to specialized farms. Specialization ranges from cultivated fruit, vegetable farms to the vast stretch of wheat and corn fields with usage of modern laborsaving machinery like tractors, combined harvesters to air borne spraying techniques. Since agricultural transformation is based on specialization, most of the specialized farms lay emphasis on monocrop cultivation, usage of capital-intensive technology, and relying on economies of scale to cut back costs of production and focus on profit maximization. Agricultural transformation in a way is more similar to operations carried out in large industrial enterprises. In both developed and developing nations, the specialized farming operations are managed by multinational corporate enterprises with large scale agribusiness activities.

Marketing of Agricultural Produce

The process of agricultural transformation towards marketization depends on the choice of objective. In case of big farms, market-orientation is not a big problem. But small farmers have to face a number of difficulties in marketing their produce under the framework of agricultural transformation. The farmers of small landholdings are usually indebted and in order to repay their debt obligations they are forced to sell off their produce at lower prices immediately after harvesting. The small farmers do not have the option to store their agricultural products to sell at a later date for the right price. On the contrary, these small farmers become the victims of the middlemen. Further, due to lack of proper infrastructure and transport facilities, they fall back to the middlemen. Hence, farmers although practising agricultural transformation, i.e., going for crop specialization with an eye on the quality and the quantity, cannot enjoy higher prices in nearby trading centers or towns.

The scenario is a bit different for the large farmers. On an average they can manage their own marketing system since they are not very much indebted, can stock the product, and have high bargaining power.

In the small farm model, a proper marketization requires some initiatives on the part of the government. These can be described as under:

- Firstly, small farmers should have the option to store their products.
- Secondly, introducing minimum support price should be encouraged so that small farmers are not forced to sell their products immediately after harvesting.
- Thirdly, the middlemen or the fariah should be removed as the linkage between the agricultural producers and the market. Here, cooperative marketing societies can act as a channel to sell their agricultural products.
- Next, the trading centers and the remote villages should be interconnected with proper transport facilities.
- Fifthly, there should be ample number of warehouses and cold storage houses for storage and preservation of agricultural produce.
- Sixthly, regulated markets need to be established for an efficient marketization under agricultural transformation.
- Seventhly, a very important aspect of marketization is availability of cheap credit. Easy and economical supply of credit helps the farmers to reap the advantages of agricultural transformation.
- Finally, a proper network market information system makes the farmers aware of the current market prices.

Marketing Information System

Agricultural transformation is basically commercial and market-oriented, focusing on profit maximization as well as vigilance on the quality and quantity of the produce. Maintaining market-orientation under the framework of agricultural transformation requires proper network of market information. The main objective of the Market Information System (MIS) is to encourage the decision-making process, mainly of the small scale farmers. The gamut of this MIS is aptly designed with well-defined responsibility and procedures with various means for collecting information covering reports and records, market research, market intelligence system, and different models of market information. The role of informal network and information sharing is also of considerable importance. In this entire process of agricultural transformation, dissemination of information benefits the farmers, agricultural entrepreneurs and the entire system. The MIS supports face-to-face communication for information collection, analysis, and dissemination and also encourages organizing training workshops, trade shows, and other events.

The marketing information system is designed in varied ways for different economies with varying usage of technology and communication media, components of information collection, and the type of users.

In agricultural transformation, emphasis is put on agribusiness enterprises to support the decision-making process and restructure the marketing process. Market-orientation is a part of the agricultural transformation process and the MIS is internally funded. Agricultural transformation can meet with only partial success in absence of proper marketing information and support mechanism. Since lack of marketability of agricultural produce, the farmers are forced to sell their crops to middlemen. These middlemen block market information creating situations in their favor and enjoying a large chunk of profit from agricultural tradable. The poor farmers are at a disadvantaged position, unable to gather the correct information on the right price, demand for the agricultural products, and alternative channels of marketing.

An illustration of the market information system can be put up in the context of Indian Agriculture. Over 70% of the Indian population is confined in villages with the poor farmers just owning only a hectare of land. They are devoid of the current status of the agricultural market and how their farming activities are going to influence the sales. With a feeble rural, social and institutional set up in India, the poor farmers exercise little bargaining power. A stepping stone was made in this regard by ITC (Indian Tobacco Company), trying to diversify itself from the shrinking tobacco industry. A new business model was established by associating the rural farmers with the market information system and support mechanism. In rural areas, Choupals were considered to be a meeting place and was selected as the entry point for the ITC. To address the problems of fragmented and dispersed agricultural markets, ITC came up with the concept of e-Choupal. This innovative idea offered direct business opportunities with ITC, furnishing market information about the agricultural products and allowing the farmers to explore the competitive avenues. In this business model, ITC staged Internet kiosks and converted them into e-Choupals. Even the remotest villages, which lacked electricity, were enabled with VSAT satellite and solar batteries. Farmers were even trained to be market-oriented and expert within the system. The e-Choupal was connected to tailor-made (by ITC itself) websites for the targeted farmers to update them about the marketing aspects of certain cash crops like soya, wheat, coffee, and shrimp. Facilities were provided even for input procurement like seed, chemicals, and other fertilizers. The farmers were helped by an e-Choupal manager known as the 'Sanchalak' to use the different agriculture-specific crop sites.

E-Choupal helped the process of agricultural transformation towards market-orientation. The farmers gathered knowledge about the agricultural market, the price trends, farming techniques, links with different agricultural universities, and credit sections of different financial institutions, meteorological departments for weather forecasting, even selling agricultural produce and procuring raw materials online. Since under agricultural transformation, the focus is on crop specialization, emails provided them (specialized farmers) to contact agronomist for various crop-related queries and finally availing services like crop insurance, sale and hire tractors, soil testing and harvesters. Therefore e-Choupal links up the Indian farmer with the consumers in agricultural markets in matters of marketing and distribution of agricultural commodities.

Figure 1: ITC e-Choupals Marketing Information System Framework

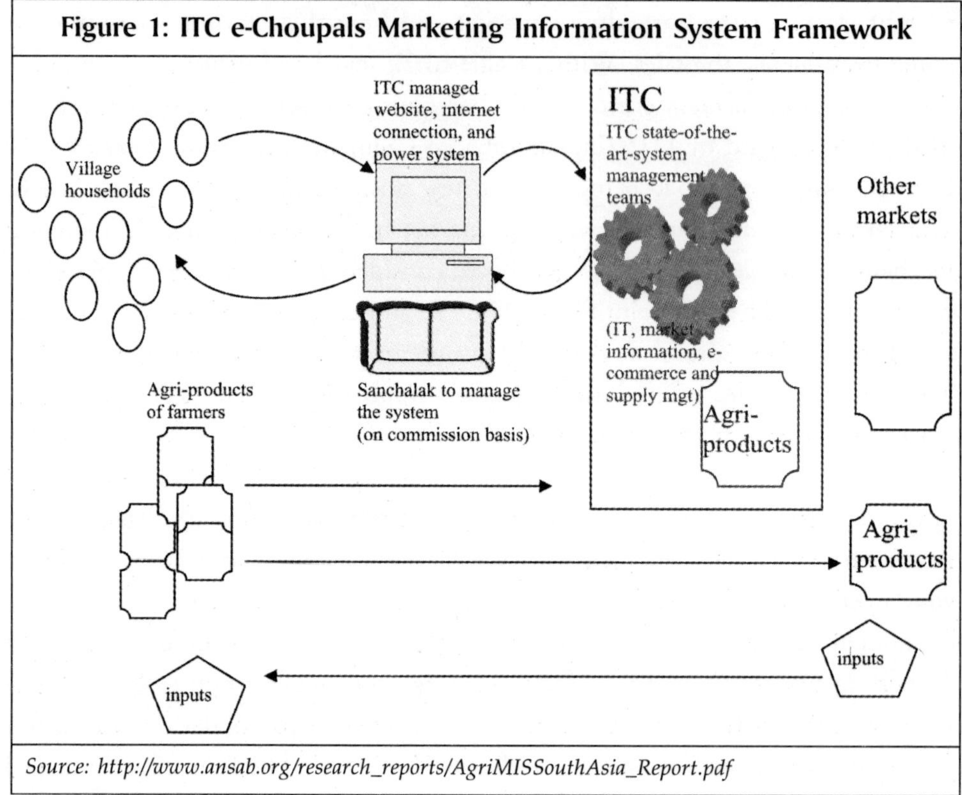

Source: http://www.ansab.org/research_reports/AgriMISSouthAsia_Report.pdf

Conclusion

The benefits of agricultural transformation will be incomplete in absence of market-orientation. Agricultural transformation is mainly a process by which individual farms swing from highly diversified subsistence oriented production towards a specialized production process aimed for the market or other systems of exchange like long-term contracts. In a way agricultural transformation is relying on the "JIT" (Just-in-Time) framework focusing on input and output delivery systems with a greater integration of agriculture with other sector of domestic and international economy. Agricultural transformation is a subset of structural transformation, the latter emphasizing on generation of economic output and employment by sectors other than agriculture.

(Nirbachita Karmakar, Senior Research Associate, Icfai Research Centre, Kolkata. The author can be reached at nirbachitak@iupindia.org).

References

1. *http://www.aec.msu.edu/fs2/ag_transformation/Def_Trans.htm*
2. *http://www.ansab.org/research_reports/AgriMISSouthAsia_Report.pdf*
3. *http://qed.econ.queensu.ca/pub/faculty/lloyd-ellis/econ239/readings/todaro9.pdf*
4. *http://mpra.ub.uni-muenchen.de/7771/1/MPRA_paper_7771.pdf*
5. *http://www.ier.hit-u.ac.jp/~kurosaki/punjab2.pdf*
6. Sarkhel Jaydeb, (1990), Economics, Book Syndicate.

5

Ecosystem Approaches to Research on Agricultural Transformation, Nutrition and Human Health

The article underlines the agricultural transformation playing an important role in creating supportive conditions for human health as many aspects of farming practices influence both rural and urban populations' health. Currently, in many developing countries, agro-ecosystem management practices are in a process of transformation due to many reasons. The primary objective of ecosystem approaches on agricultural transformation and human health is to deepen the understanding of the relationships between agricultural transformation and human health with a specific emphasis on food security, dietary diversity and their implications on nutrition.

Underweight and nutrient deficiencies together represent the leading risk factors in developing countries contributing to deaths and the burden of disease.

– World Health Report: Reducing risks, promoting healthy life, WHO, 2002.

Human health and well-being are dependent upon various factors, ranging from economic aspects of a society to the quality of the environment. Agricultural practices, in particular, play an important role in creating supportive conditions for human health as many aspects of farming practices influence both rural and urban populations' health.

Currently, in many developing countries, agro-ecosystem management practices are in a process of transformation due to increasing population pressure, decreasing availability of agricultural lands, environmental constraints (such as climate change), contamination from industrial and agricultural activities, increasing urban market demands and globalization processes. Agricultural transformations,[1] especially, when coupled with economic and environmental degradation, can adversely impact the relationships between people and the ecosystems on which they depend for their livelihood, affecting patterns of human health, disease and nutritional status. For example, a shift away from subsistence agriculture and an increasing orientation to markets for both income and food purchase are major changes that affect local systems of production. While more intensive (commercial) agriculture can offer economic benefits to rural populations and reduced food costs for consumers, it has mixed impacts on nutritional status, in part because of the erosion of local crops and varieties that underpin traditional dietary diversity. Changes in land use, including disturbance, deforestation and appropriation of natural areas, diminish opportunities for hunting and gathering the essential wild components of many traditional food systems. In addition, some niches or farming practices such as home-gardens assume less importance even though they provide complementary resources for diet and medicine and also serve as repositories of biodiversity. Common detrimental health effects of these changes include insufficient food and impoverished diets, leading to malnutrition, weakened immune systems and more frequent infections, often of higher severity and longer duration, as well as an increase in diabetes, heart disease and obesity.

In developing countries, agricultural programs and subsidies have generally focused on food security, attempting to increase food availability for populations showing overall calorific uptakes deficit. These efforts often led to a simplification

1 Agricultural transformation can mean a range of changes in food production practices, such as the intensified use of inputs or the introduction of new practices and technologies.

of diets to a limited number of high-energy foods.[2] While this issue is indeed important, research demonstrates that it is equally critical to address dietary diversity, non-cultivated foods, micro-nutrient research with food-based & non-food based approaches, as well as protein-malnutrition which may also have considerable impacts on human health in both developed and developing countries. Moreover, there are multiple socio-ecological determinants that influence the nutritional quality and quantity of individual needs, which should also be considered in research projects on nutrition and human health. Access to quality food is often an issue of affordability rather than availability that emphasizes the importance of poverty reduction considerations. The challenge is how to address a problem whose causes and consequences span culture, health, agriculture, markets, and environment.

The aim of this year's Ecosystem Approaches to Human Health Awards Program is to deepen our understanding of the relationships between agricultural transformation and human health with a specific emphasis on food security, dietary diversity and their implications on nutrition. Relationships between local food production practices and household nutrition are multiple and addressing them represents a pertinent strategy for enhancing the health of disadvantaged populations in developing countries. Interested applicants for this year's Ecohealth Awards Program are encouraged to submit proposals on two particular aspects of nutrition research: food security and dietary diversity.

Food Security

For the last 50 years, food security has been a key issue of concern for international agencies striving to support programs for better health of the poor populations of Africa, Asia, and Latin America as food availability is considered to be one of the principal determinants of health status. Even though numerous programs have been designed and implemented to increase food security in the world, meeting the dietary needs of vulnerable populations still remains a challenge in 2005.

The concept of food security goes beyond the simple consideration of food quantity; achieving food security implies the availability of sufficient food of

2 The "nutrition transition" is characterized by the simplification of diets thus reducing the consumption of diverse, nutritionally rich, and functional plant foods and increasing the consumption of energy-rich foods that are nutritionally poor and that contain fewer functional properties for human health.

good quality for every individual of a given population. The complex nature of food security indicates that in order to achieve a food secure status in a population, a range of aspects need to be assessed, such as equitable sharing among household members, sufficient household production dedicated towards subsistence, use of soil, water and biodiversity conservation techniques, and the sanitary aspects of cooking. The study of food security requires a careful exploration of socio-ecological determinants that modulate individual and community access to food for optimal health for all members of a community.

Dietary Diversity

Dietary diversity is commonly recognized as a key component of high quality diets. Its relevance to disadvantaged populations in developing countries stems from the propensity for persistent nutrient deficient diets and the importance of increasing both food and food group variety to ensure nutrient adequacy. Beyond this, there is increasing recognition on the value of variety in food functionality (non-nutrient properties of food). Many of the benefits of non-nutrients are very important and have longstanding traditional reputations and use in many parts of the world. Examples of this include fiber, fat and phytochemicals (e.g., antioxidants, anti-inflammatory constituents, immunostimulants, and others).

It is important to consider the social and ecological determinants of dietary diversity because a diversifying diet may take different forms depending on the particular context. In some cases, introducing or reintroducing different meat products in the diet can enhance protein and iron uptake, often scarce in cereal-based diets. In other cases, vegetables and fruit uptake should be promoted in order to provide supplemental sources of micro-nutrients. Diet diversification may be linked to hunting and gathering or to the cultivation and domestication of wild foods (plants and animals), depending on the local specificities. Traditional knowledge is an important resource to assess the value of food that was once part of the diet.

Ecosystem Approaches to Human Health (Ecohealth)

Ecosystem approaches in this Awards competition can be seen as useful processes to deepen our understanding of the linkages between human health, nutrition and agricultural transformations. The approaches recognize that there are inextricable links between humans and their biophysical, social, and economic environments that are reflected in individual/communities' health. They focus on understanding (i) the interactions between social and ecological systems in defining key determinants of human health in particular settings, and (ii) the impact of human activities on the sustainability of these processes. They also seek to identify ecosystem management strategies that contribute to improving the health and living conditions of human populations and the sustainability of the ecosystem in which they live. Ecohealth represents a process-oriented and dynamic way of understanding and solving problems, which can be constructed in various contexts, with varying scales, and different intended outcomes. "Ecosystems' in this approach are defined relative to the research problem and refer to the social and ecological contexts, both on a temporal and a spatial scale, of human lives. Human activities (or stressors) alter these contexts and have positive and negative effects on individuals and communities involved.

The Ecohealth approach currently has three core elements or pillars: i) trandisciplinarity; ii) social justice and gender equity; and iii) multi-stakeholder participation. These elements are key to improving health and well-being as they allow for an understanding of change that explicitly link social and ecological systems. Applications for this year's Awards should clearly detail how the proposed research will integrate the three core elements of Ecosystem approaches to human health.

Transdisciplinarity

Ecohealth research project have taken the premise that local food production systems, food availability and food diversity are intrinsically linked and influence the health status of human population. They should, therefore, be addressed together in a research project from a human health perspective in order to elaborate strategies that can improve the health and well-being of disadvantaged populations in developing countries and the sustainability of the ecosystem on which they depend for their livelihood. The complexity of relations between these processes clearly stresses the need for transdisciplinarity in project design and implementation.

Transdisciplinarity is a process through which researchers from different disciplines and stakeholders involved in a research project (policymakers, community members, civil society organizations) transcend the limits of their own disciplinary background and knowledge to contribute to the development of new concepts, hypothesis and knowledge in order to jointly develop and implement potential intervention strategies. In a research project transdisciplinarity implies that all stakeholders contribute to the conception of the project, the identification of the research questions and collaborate in the realization of the research. Stakeholders must meet regularly throughout the project to discuss their work, its implication for the research and, if necessary, to reassess the research questions, hypotheses or activities.

For an academic research supported through the Ecohealth Awards Program, it is clear that students cannot cover the different disciplines relevant within their own research. Applicants should contextualize their research project and clearly explain how it contributes to the better understanding of the complex linkages between socio-economic and environmental factors and their influence on human health. Strategies on accessing expertise outside the student's own capabilities should be a critical part of a proposal submitted for the Ecohealth Awards Program

Stakeholder Participation

The involvement of multiple stakeholders is fundamental for the success of a project using an ecosystem approach to human health. The approach emphasizes the need to consider two levels of participation: the participation of local communities and the participation of other stakeholders such as policymakers, healthcare providers, agricultural extension workers and civil society organizations. Aiming to improve nutrition and human health in relation to local food production practices clearly requires the involvement of community members but the sustainability and scaling up of potential interventions are conditional on the involvement of several other stakeholders. It is essential to maximize a sense of ownership for the project and to involve stakeholders at all stages of the research.

Gender and Equity

The Ecohealth approach also emphasizes the need to consider social and gender equity in the quest for new knowledge and interventions to improve human health and well-being through better natural resources management. Gender is especially relevant for research targeting health and nutrition issues since men and women assume different responsibilities within a household, have different levels of influences on decisions and control over resources. The distribution of roles and responsibilities between selected members in a household can influence their health and nutritional status, vulnerability and capacity to react to health and nutritional problems. Gender requires going beyond sex disaggregation (the description of how men and women may be affected differently by specific problems) to consider how power relations and differential access to, or control over resources influence the vulnerability of men and women to health problems.

Men and women, however, are not homogenous groups. Gender is one among several social divides influencing the vulnerability of specific groups in a community that must be addressed by the research proposals submitted to the Ecohealth Award Program in order to identify the most vulnerable groups. It is essential to understand the impact of gender and social relations on health and nutritional issues to identify potential entry point for intervention.

Agricultural Transformation, Nutrition and Human Health – The Linkage

– Subir Ghosh

Agricultural transformation is a process through which a single farm shifts from traditional production system to a highly specialized production system towards market orientation. In other words, it involves a range of tools encompassing traditional techniques to improve domesticated plants for enhancing their qualities with regard to production efficiency or their end use and characteristics. It has often been proposed as a supplementary strategy to combat micronutrient malnutrition. Golden Rice (GR) could be used in intermediate strategies to address vitamin A deficiency, one of the prevailing micronutrient deficiencies in many resource poor countries. Proactive and targeted research-based implementation and broad public support can bring the potential of agricultural biotechnology to its optimum fruition, through its contribution to a sustainable as well as permanent solution of nutritional disorders. For over a decade now, agricultural biotechnology has demonstrated its potential to play a role in addressing global food crisis by raising the production level.

Globally, one of the most prevalent but preventable nutritional problems today is micronutrient deficiency due essentially to insufficient levels of dietary vitamins and minerals. Micronutrient deficiency is the most common type of malnutrition. It diminishes cognitive, physical, and reproductive development. Along with most low-income countries, some middle-income countries like China also suffer from the menace of micronutrient deficiencies.

Rice is the staple food in many parts of developing world, producing geranyl diphosphate – a precursor of beta-carotene. But the nutritive value of the commonly consumed rice is rather poor. Besides, VAD (Vitamin A Deficiency) affects hundreds of millions of children worldwide, thereby causing blindness in many more children every year, many of whom die within a year of losing their sight. Scientists have been working on A major guiding force is to address the two major health challenges malnutrition and deficiency of Vitamin A, both of which have severe consequences on health and life expectancy. This particularly holds true in most developing countries, where rice is the staple food (Pinstrup-Andersen and Cohen 2001).

(Subir Ghosh, Senior Faculty Associate, Icfai Research Centre, Kolkata. He can be reached at subirg@iupindia.org).

References

Hussain Azmal (2008), "Agricultural Biotechnology in Combating Nutritional Disorders Current Perspectives" published in the book *Agriculture Health Linkage: Emerging Issues*, The Icfai University Press.

Pinstrup-Andersen, P and M J Cohen (2001), "Modern Agricultural Biotechnology and Developing Country Food Security", in: Nelson, G. (ed.), *Genetically modified organisms in agriculture*, San Diego *et al.*, 179-189.

6

Agricultural Transformation and Changing Status of Women

P Narasimha Rao and R Venkata Rao

This paper highlights that agriculture provides sustainable livelihood to the rural poor and would improve their socio-economic status and norms. Women play an important role in the agricultural sector in the production, processing, storage and marketing activities, in addition to their valuable services in the domestic sector. This study examines the economic, social and political role of women and the changing position of rural women due to the technical changes and agricultural development in the process of liberalization of Indian economy. It points out that agricultural growth had both negative and positive effects on women from different socio-economic strata. The agricultural growth has indicated an improvement in the status of upper socio-economic strata on one hand, and deterioration in the status of lower socio-economic strata on the other. Therefore, the rural transformation consequent to agricultural growth enabled the women to have access to education and productive resources. The agricultural growth

Source: The Icfai Journal of Agricultural Economics, April 2005. *This article also appeared in the book "Empowerment of Rural Women: Insights and Initiatives" published by the Icfai University Press.*

has not only led to a decline in infant mortality rate and maternal mortality rate, but also led to changes in the pattern of gender gap and literacy among women. Further, this study concludes that agricultural transformation brought changes in the health and marriage practices of women as well as political participation among women.

The globalization of economic life goes beyond the boundaries of narrow trade issues and its suspected impact on rural people, particularly women, is expected to be significant. The rapidity of international capital transfers and widespread utilization of information technology all act together in generating a new environment for development. While globalization provides significant new opportunities for improving rural livelihood, (opening of new markets, access to new technologies, improved social service coverage, etc.) it can also present significant threats, since its benefits can easily bypass those who are isolated and marginal, like women and other disadvantaged members of the society.

Agriculture has been the main source of livelihood for millions of Indians living in rural areas. Agricultural development brings about sustainable livelihood to the rural poor and would improve the socio-economic and political status and norms. Women play an important role in the agricultural sector. Their role in production, processing, storage and marketing is well-established. In addition to this, their contribution to the work relating to socialization for children and domestic work cannot be quantified. Agricultural growth consequent upon the globalization of Indian economy altered the organization of production, patterns on ownership of land and cropping patterns, which in turn, altered the structure of inequalities in rural economy. This change in the technology and organization of production altered the multiple roles of women and consequently, the status of women in the family and society. In this context, it is appropriate to study the economic, social and political roles and changing position of rural women consequent upon the technical changes and agricultural development in the process of globalization of Indian economy.

The agricultural growth had both negative and positive impact on women from different socio-economic strata. On one hand, it relieved women belonging to the upper socio-economic strata as they have withdrawn from the framework and were utilizing their free time in other areas; here a clear trend of improvement in their status is evident. On the other hand, it had negatively affected women from lower castes and lower income groups.

Irrigation, which is an important component of agricultural transformation has also affected the role and status of women. There is a shift from barter to monetized economy along with agricultural transformation. The workload of women in post-harvest operations has increased and their work on irrigated crops further increased while their workload on food processing has decreased. Irrigation has increased the wage work and has stopped emigration. It has also solved the problem of women in procuring drinking water to some extent and has altered the role of women in household too. Animal husbandry has become a women enterprise. The average family income went up substantially. Modernization of agricultural sector is expected to alter the access of women to control liquid assets. Rural transformation consequent upon agricultural growth will enable the women to have access to education, training facilities and productive resources at various levels.

But in the Indian context, planned development is initiated and implemented through the existing institutional structures. These institutions have practical agenda bias, with meager representation of women in their decision-making. In this situation, technological transformation and agricultural development may have an adverse impact on the status of women. As a result, agricultural modernization might have strengthened only the existing institutional arrangements thereby making women economically deprived, socially segregated and politically a powerless group.

Women play socially and economically productive roles but their social and productive roles are governed by values, norms and customs of the society, which is seen to have placed them in the lower social and economic status. The role and status of women are altered in the process of agrarian social transformation due to the technology of cultivation. The position of women, both in the family and in society has radically changed. The disaggregated pattern of development has varied impact on women members of every region, group, class and caste. In this backdrop, the present study proposes to examine the status of women across the levels of rural transformation consequent upon the technological and agricultural growth in the Guntur district of Andhra Pradesh in India.

Rural transformation, led by agricultural development which resulted in social change, has had its effect on the status of women across regions with varied levels of development. So the area of study is divided into zones according to its level of transformation and agricultural development. It is evident that the agricultural modernization and rural transformation is associated with irrigation. Agricultural development in Andhra Pradesh mostly took place with the irrigation infrastructure. So, the district is divided into a Highly Developed Zone, Medium Developed Zone and Less Developed Zone, on the basis of levels of irrigation viz., highly irrigated, medium irrigated and rain-fed areas. Areas having irrigation for two crops is termed a Highly Developed Zone and areas having irrigation for one crop is termed as Medium Developed Zone and areas where agriculture depends upon rainfall as Less Developed Zone. For the purpose of the study, three villages from each zone have been selected at random by using multistage random sample.

Data for the study is collected by village census and household survey through structured intensive schedule. Village census was undertaken to collect information about the demographic, social, economic and occupational aspects, and the household schedule is used to collect information about the status of women from sample households.

To examine the broad social process generated due to the new technology used in cultivation and its impact on the changing status of women within the existing social process, the following indicators have been taken.

The sex ratio in the population, infant mortality, literacy, age of marriage, political participation, decision-making process both in household and political arenas, access to assets and access to medical facilities are taken as indicators to examine the changing role and status of women in the process of technological, economical and social development as a result of changes in the agricultural sector.

The most important indicator of the status of women in any society is their numerical strength. Sex ratio reveals the relative status of the number of women in a society and indicates various issues involved in the gender imbalance. Table 1 reveals the data on sex ratio in the study zone.

Table 1: Female Sex Ratio

Zones	Female for 100 Males
I	89
II	98
III	94

Source: Sample Survey.

Table 1 reveals that there is a significant variation in the three zones. Highly developed zone shows lowest sex ratio, Zone-II shows equal sex ratio and Zone-III shows state trend. Although Table 1 reveals existing female sex ratio in the study zones, it doesn't reflect the long-term trend. To examine the long-term trend of female sex ratio, the census data for the last 40 years has to be examined. To examine the impact of agricultural transformation and changing sex ratio, the census data from 1961 is given in Table 2.

Table 2 clearly reveals that the female sex ratio is higher in Zone-I when compared to the other two zones and the same is higher in Zone-II when compared to Zone-III, i.e., female sex ratio is directly related to the levels of agricultural development. This may be due to the fact that female infant mortality rate and maternal mortality is decreasing with the access to medical facilities consequent upon the increase in the levels of income and sustainable employment.

Table 2: Sex Ratio

Year	Zone-1	Zone-2	Zone-3
1961	96	96	84
1971	97	97	84
1981	98	97	91
1991	97	96	92
2001	98	96	91

Source: Census.

Table 3 reveals that infant mortality rate is higher in Zone-III and lower in Zone-I, i.e., the agricultural development along with irrigation has led to rural transformation where people have more access to education and medical facilities, i.e., infant mortality rate is decreasing with the increasing levels of income and agricultural growth.

Table 3

Zone	Infant Mortality Rate	Female Infant Mortality Rate
I	60	32
II	68	39
III	74	48

Source: Sample Survey.

High female infant mortality rate is observed in Zone-III, when compared to Zones-I and II. This clearly reveals that the people in Zone-III do not have access to education and medical facilities because of low income and poverty when compared to other zones where agricultural development has contributed to rural transformation.

Literacy and Education

Illiteracy is the root cause for all the problems in rural India. Literacy rate is very low in rural areas and literacy among women is still lower when compared to men.

It is very low among Scheduled Castes and Scheduled Tribes. Table 4 reveals female literacy rate in these zones. Female literacy rate is very low in Zone-III where adoption of agricultural technology and agricultural growth is very low. Female literacy rate is directly related to levels of adoption of new agricultural technology and development. The transformation of agricultural sector enable the rural households to increase their income sufficiently whereby they can afford to send their children to schools. This has contributed to high literacy rate among women in agriculturally developed zones.

Table 4: Female Literacy Rates

Zone	Male	Female	Total
Zone-I	78.81	67.87	73.35
Zone-II	74.27	55.24	64.86
Zone-III	49.35	24.36	37.02

Source: Census, Survey by Schedule.

Comparing the data with the census of 1971-2001, the table indicates that women's literacy has become more than double in Zones-II and III while it is three times more in Zone-I, which has developed well by agricultural transformation. Table 5 reveals a substantial improvement in women's literacy rate since 1971. Women's literacy growth rate is higher in Zone-I when compared to the other two zones. This shows that rural transformation consequent upon the adoption of new agricultural technology and agriculture development might have contributed to the growth of literacy rate.

Table 5: Changing Women's Literacy Rate

Year	Zone-I	Zone-II	Zone-III
1971	20.4	20	11
1981	43.8	23	16
1991	60.6	41	22
2000	69.0	55.71	34.36

Source: Sample Survey.

There are divergent views about the impact of agricultural growth and rural transformation on the income distribution of the rural households. The agrarian society has witnessed diverse levels of technological transformation, economic development and social change. This disaggregated pattern of development had a varied impact on women members of every class and caste of the agrarian society in general and on the status of women in particular.

In recent years, there have been significant changes in the pattern of gender gap in literacy among social groups due to the technological transformation. Table 6 reveals the gender gap in literacy among different social groups. The gender difference in Zone-III is very high when compared to the other two zones. Similarly, gender gap is less in Zone-II when compared to Zone-III; this reveals that gender gap in

Table 6: Percentage of Literacy by Sex and Caste ***(in %)***

Caste Groups	I			II			III		
	Total	Male	Female	Total	Male	Female	Total	Male	Female
Upper castes	54	65.9	37.0	48.2	55.3	26.8	42	56.7	22.0
S.C.	42	45.0	34.9	18.2	23.6	9.8	29	25.0	11.2
S.T.	26	30.2	19.2	25.1	32.0	8.9	16	22.0	9.0
O.B.C.	67	8.0	52.3	45.0	63.2	31.3	38	45.0	16.2
Minorities	64	56.0	49.2	55.0	52.0	11.2	42	54.0	21.0
Total	48	56.9	36.5	36.0	3.9	16.0	29	26.0	12.3

Source: Sample Survey.

literacy has been reduced by technological transformation and resultant agricultural development. And it is evident that gender gap in literacy is low in upper castes in all the zones. It is clearly evident from Table 6 that gender gap in literacy is directly related to social hierarchy and also agricultural development.

The variation in gender gap in literacy is due to the variations in investment on education. The higher expenditure on education may have a direct impact on economic development and occupational diversification, but the higher the level of agricultural development, higher is the level of investment on education which might have reduced the gender gap in literacy.

Access to Healthcare

The access to healthcare facilities for women and the quality of these facilities depend upon availability of services within a reasonable distance at an affordable cost and the economic conditions of the family. Though Government institutions are available, women are using indigenous systems of treatment and private doctors. Though facilities are available in the private sector where quality is also available, it is not within the reach of majority of the women. For common diseases like fever, headache, stomach disorder, burns and other injuries, the healthcare practices of women depend on their income and economic status. Table 7 reveals that only economic empowerment, i.e., sustainable income enables women to use modern health practices.

In case of serious maternity problems 68 percent of women in Zone-I, 42 per cent of women in Zone-II and 28 percent of women in Zone-III went to qualified

Table 7: Healthcare Practices for Common Diseases *(in %)*			
Zone	**Indigenous**	**PHC**	**Private Doctors**
Zone-I	49.7	5.4	44.6
Zone-II	52.6	19.2	28.2
Zone-III	68.4	26.2	5.4
Source: Sample Survey.			

doctors and the remaining depended on home treatment and local midwives. But these women never went for prenatal checkups. Even now childbirths are taking place at homes, though Government hospitals are accessible. The maternal mortality rate is 5.2 per cent in Zone-III, 4.2 per cent in Zone-II and it was not estimated for Zone-I during the year of survey. Though health facilities are created by the Government in rural areas, they are not accessible to women because of constraints like lack of financial resources, technical staff and inadequate facilities in subcenters, Primary Health Centers and Community Health Centers. So, it is clear that women empowered economically through agricultural transformation and increased incomes have access to the health facilities in private sector.

Marriage Practices

In the Institution of Marriage and marital status, degree of consent or choice expressed, exchange of dowry, traditional rituals and customs are the key indicators of the relative status or degree of subordination of women in the society. The marriage practices are mostly guided by customary norms, values and beliefs of the caste and community. Marriage is considered to be sacred in the Indian society. But the process of technical transformation, economic development and occupational diversification brought about changes in the age of marriage and marriage practices.

One of the key indicators for social development of women in rural areas is the mean age at marriage. Mean age at marriage in a society is a mere function of cultural factors in rural areas where literacy among women has a positive influence. It is a fact that mean age at marriage will not increase unless efforts are made by creating awareness and female literacy. It is evident from Table 8 that literacy is positively correlated to the economic development. The inequalities in economic development have been responsible for variation in mean age at marriage. Table 8 reveals that higher mean age at marriage among women is brought by

Table 8: Mean Age at Marriage

Zone	Below 14		15 to 18		19 to 21		22+	
	Male	Female	Male	Female	Male	Female	Male	Female
I	1.2	22.0	24.3	34.0	32.4	32.0	42.1	12.0
II	5.4	24.0	25.1	40.8	34.2	30.2	35.3	5.0
III	9.0	25.0	34.2	42.2	36.7	28.3	20.1	4.5

Source: Sample Survey.

commercialization of agriculture, spread of education and communication in developed zones. Child marriages are on the decline and negligible in Zone-I wherein agricultural transformation took place. If they exist at all, they are confined to socially deprived sections of the society.

Choice of Marriage Partner

The process of economic development, modernization and social change brought about changes in the marriage practices. Table 9 reveals the marriage practices relating to choice, dowry and degree of village exogamy in these zones. It can be observed that in the choice of partner there is no clear relationship with modernization and its influence. The trend is clearly in favor of some form of consent to matches arranged by the parents. Self-arranged marriages are practiced more in less developed areas than agriculturally developed and semi-developed zones. In Zone-I and Zone-III, the menace of dowry is high. Parents preferred to arrange a suitable match with minimum dowry than the girls' consent. Economic development and occupational diversification has contributed to the rise of dowry system. The penetration of cash economy has brought the evil of dowry into the rural areas.

The distance of matrimonial homes indicates the extent of family, caste and kinship. Most of the women in developed zones married in adjacent villages and outside the district. But in Zones-II and III most of the women married within the village and adjacent villages.

Table 9: Marriage Practice Adopted for Women *(in %)*

Zone	Matched/ Arranged by Self	Arranged by Parents	Dowry Given	Within Villages	Adjacent Villages	Outside District
Zone-I	97.3	92.7	80.5	18.4	57.0	25.0
Zone-II	4.8	95.2	50.2	31.2	61.0	15.0
Zone-III	4.6	95.6	62.6	15.2	66.2	04.2

Source: Sample Survey.

Political Participation

Political rights exercised by women is an indicator for improving their status in Indian Society. Political participation of women in rural areas is confined to electoral processes in the *Gram Panchayats.* Table 10 reveals about the political participation of women. Though there has not been marked differences in voting participation, the pattern and reasons for their participation has not been same in all the zones with various levels of agricultural transformation. The number of women actively joining in the political campaign and meetings has been very low in less developed areas. Dependence on political decision-making is more in the less developed zones.

Table 10: Political Participation of Women *(in %)*

Zone	Active in Political Campaign	Active in Political Meetings	Casting Votes	Self-determined	Decision by Relatives
I	8	12	96	39.4	38.3
II	5	02	98	24.2	50.0
III	–	01	97	22.2	62.0

Source: Sample Survey.

Conclusion

Irrigation has become an important component of agricultural development. Agricultural development transformed the rural economy and has affected the role and status of women. The rural transformation consequent upon agricultural growth enabled the women to have access to education and productive resources. The status of a number of women increased. Agricultural growth has led to a decline in the infant mortality and maternal mortality rate. Agricultural development has also led to changes in the pattern of gender gap in literacy among social groups. Gender gap in literacy is directly related to social hierarchy and agricultural development. Agricultural transformation brought about changes in the health practices, marriage practices and also political participation among women.

Policy Implications

In spite of significant achievements in areas such as per capita income, poverty reduction and literacy of women, they have been confined to the agriculturally developed regions. The regions particularly where the mode of agricultural production is traditional and mostly depending on rainfall have made slow progress in improving the socio-economic conditions of women.

There is an urgent need to improve the investment climate in regions where agriculture is traditional, to raise the periodicity by tackling the infrastructure bottlenecks in irrigation, power and suitable dry land agricultural technologies. The policy process should emphasize the urgent need for reforms in dry land or unirrigated agriculture which is critical for development of disadvantaged members of the society like women and poor. The policy should aim at enhancing the opportunities for improving the status of women which requires investment in human development, that is, access to quality of expenditure on social services such as health, education and access to new technology. This requires a major reorientation of current programs to suit the local needs of backward rural areas.

It is a fact that the development needs of women are diverse and multifaceted. In order to address these diverse needs better, specific development approaches should be followed at national, regional, and community and household levels. It is also realized that the small rural women producers have the potential to become successful economic actors. But in order to translate their potential into practice, there must be a clear focus on a number of areas.

The policy mechanism should ensure women, equitable access to resources like natural, financial, human and representation; through participatory tools and process they should be given clear incentives so as to successfully mobilize their own resources. It is only literacy that enables them to manage risk and minimize vulnerabilities. The policy process should focus on how to improve the profitability of economic opportunities in the agricultural sector where women play an active role by giving the rural poor a voice through participation, access, training and communication and maintaining a gender focus.

(P Narasimha Rao, Associate Professor, Department of Economics, Acharya Nagarjuna University, Guntur. He can be reached at pemmasaninrao@yahoo.com,

R Venkata Rao, Assistant Professor, Department of Economics, Acharya Nagarjuna University, Guntur, Andhra Pradesh.)

References

1. Awasty L, (1982). *Rural Women in India: A Socio-Economic Profile of Jammu Women,* New Delhi: B.R. Publishing Corporation.
2. Bhaduri A, (1985). "Technological Change and Rural Women: A Conceptual Analysis in Technology and Rural Women" in Ahmad I, (ed.) London: *Technology and Rural Women,* George Allen and Unwin.
3. Law P, (1985). "Rural Development: It's Impact on Women", *Social Action,* 35(1) pp. 80-90.
4. Mazumdar V, (1983). *Women and Rural Transformation,* New Delhi: Concept Publishers.
5. Mukherjee P N and Chattopadhyay M, (1981). "Agrarian Structure, Proletarianization and Social Mobilization", *Sociological Bulletin,* 30(2), pp. 137-162.
6. Ramamurthy P, (1985). The Role of Women in Irrigation: A Report on a South Indian Canal System (Mimeographed).
7. Saradamoni K, (1983). "Changing Land Relations and Women" in V Mazumdar, (ed.) *Women and Rural Transformation,* New Delhi: Concept Publisher.
8. Sharma K, (1982). Strategies for Rural Development: Impact on Women, New Delhi: Center for Women's Development Studies.
9. Stanbury P, (1981). Irrigation's Impact of Socio-Economic Role of Women in a Haryana Village (Mimeographed), University of Arizona.

Women's Contribution in Agriculture: A Global View

– Subir Ghosh

Over the last few decades, the significance of women's participation in the workforce has resulted in significant transformation in the family set up and the economy in rural as well as urban life throughout the world. Since 1950s, women's participation in economic activities has been increasing rapidly. There is no exception in agricultural activities too. Women always have participated strongly in local activities. For instance, women produce 80 percent of food grain in Africa, 60 percent in Asia and 40 per cent in Latin America. Not only do women contribute in food grain production, but also help in marketing of the products. This is a very simple example of the importance of women's participation in the society.

Women constitute around half of the total adult population in India and about 77 per cent of them belong to the rural areas. The main occupation of women in rural areas is agriculture and allied activities like development of fisheries, food processing, etc. Therefore, they contribute about three-fourth of the total labour force required for such activities. In India, about 222.52 million (33.54 per cent) women are engaged in agricultural activities and 22.09 million (3.22 percent) women are engaged in other allied activities. Therefore, many of these women contribute to rural welfare activities through their association with agriculture. While the proportion of male workers in agricultural activities decreased from 70 percent to 64 percent between 1990 and 2000, the proportion of female workers remained same. It is leading to a major shift in the workers' profile in the agricultural sector, which is getting increasingly feminised. According to the findings of the survey, 73 percent of working women are engaged in the agricultural sector against 52 percent of men in India. (Dr. Arjun Sengupta, Friday, 10th August, 2007 published in *Business Line*).

In Sub-Saharan Africa, women now play major role in agricultural labour whether hired or own labour, land clearing work etc, and also manage all household activities (UNIFEM, 2002). In rural areas, both men and women take part in decision-making about what to cultivate or how to cultivate, etc. However, men have greater influence on decision-making over and above the decisions taken by their wives about cropping pattern, seed sowing and other expenditure etc. The Traditional structure of rural muslim women in Nigeria now play an active role in farm operations. They accounted for 11 per cent farm labour in 1970 whereas it was 22 per cent (either hired labour or own farm labour) in 1990 (Satio, 1992) to 29 per cent in 2000. Whether farming or non-farming activities, women also work longer hours than men. In Sub-Saharan Africa also the number of female-headed households is increasing.

(Subir Ghosh, Senior Faculty Associate, Icfai Research Centre, Kolkata. He can be reached at subirg@iupindia.org).

References

A "brief" (2007) titled: "Middle East & North Africa: Gender Overview." World Bank publication.

Ghosh Subir (2008), "Women's Participation in Rural Development", published in "Women's Participation in Rural Development: Emerging Trends in Developing Countries", The Icfai University Press.

Section II

Country Perspectives

7

Post-Communist Transformation in Bulgaria – Implications for Development of Agricultural Specialization and Farming Structures

Hrabrin Bachev

This paper incorporates a new inter-disciplinary methodology of the New Institutional and Transaction Costs Economics, and examines pace, factors and modes for post-communist agricultural specialization and farming structures development in Bulgaria. Firstly, it presents the specific Bulgarian model for farming transformation characterizing with restitution of farmland in real borders and original locations, physical distribution of assets of ancient public farms into individual shares, rapid liberalization of markets and prices, and lack of public support to agriculture. Secondly, it specifies factors for evolution of new farm structures and specialization such as badly specified and enforced property rights; big institutional, market and behavioral uncertainty; high assets specificity and dependency; lack of managerial

experience; low incentives for long-term investment; ineffective public interventions, etc. Next, it demonstrates how these factors affect organization and specialization of farming in the country explaining the evolution of a huge subsistence and part-time farming, production cooperation at a large scale, unprecedented concentration of resources in few business farms, widespread use of informal and integrated modes, etc. Fourthly, it analyzes the impact of transition on farm structures and agricultural specialization through changes in structure and share of agricultural GDP and employment, and distribution of activities between different types of farms. Finally, it clarifies efficiency and extend of specialization in dominating large business farms, production cooperatives, and numerous small-scale unregistered farms.

I. Introduction

Since the collapse of Communist system in 1989 Bulgarian agriculture has gone through an unprecedented transformation from a centrally-planed to a market-based private economy. The fundamental transition has affected significantly agricultural specialization and farming structures (Bachev and Tsuji, 2001; Bencheva, 2005; OECD, 2000). The unique post-communist 'Bulgarian experience' gives an extraordinary opportunity to study evolution and factors of agricultural specialization. However, there are no comprehensive studies on agricultural specialization during the most recent period of development.

The goal of this paper is to fill the gap and examine modes and factors of post-communist agricultural specialization and farming structures development in Bulgaria. We incorporate an inter-disciplinary methodology of the *New Institutional and Transaction Costs Economics* based on contributions of Coase (1937), Furuboth and Richter (1998), North (2000), and Williamson (1996). *Firstly,* we outline the framework for analysis of economic specialization in transitional agriculture. *Next,* we present features of Bulgarian model of farming transformation, specifying factors for development of farming structures and specialization, and impacts on agricultural specialization. *Finally,* we analyze specific governance and specialization in dominating business farms, cooperatives, and unregistered farms. The study is based on official (statistical, census, etc.) as well as original data collected from managers of 'typical' farms in all major regions. The survey was carried out in 2003 during the last Agricultural Census and covers 194 farms of different types (0,5 per cent of the commercial farms in the country). Thirty eight per cent of surveyed farms are unregistered 'individual, family, or group farms', almost twenty nine per cent are 'cooperatives', and one-third has a status of 'firm'.

II. New Institutional Economics Framework

The New Institutional Economics let us better understand 'logic' and driving factors of agricultural specialization and farming structures development (Bachev, 2004; Masten, 1991; Sporleder, 1992).

A potential to increase *productivity* gives incentives for division and specialization of labor[1]. In agricultural production specialization could be on a particular *product(s)* (crop, livestock, fruits, cereals, milk, meat-cattle, organic honey) or a specific *function(s)* (management, mechanization, plant protection, harvesting, guarding, risk taking, marketing). Specialization of labor inevitably requires *coordination* of specialized activity and *exchange* (of products, resources, rights) between individual agents. *The governance* (coordination, stimulation) of specialization and exchanges of individuals could be done by *free market* (price-movements, private negotiations), *whiting a private organization* (by a manager, collective decision-making), and/or by *a third party* (e.g., state authority).

If *transaction costs* were zero[2] then all modes of governance (market, contract, family farm, partnership, cooperative, nationwide hierarchy) would have *equal efficiency*. Then the *type* and *extend* of specialization would depend only on *technological factors* – potential to economize on production costs, explore economies of scale/scope, increase productivity, and profit from mutually beneficial exchange. However, when transaction costs are positive then the *'governance matters'* and have a significant impact on evolution of agricultural specialization and exchange. For instance, when a farmer integrates 'mechanization service' (buying

---

1 Higher productivity could be also achieved through *cooperation* of labor.

2 Transaction costs are 'costs for protection and exchange of property rights' (Furuboth and Richter, 1998: 101).

a tractor and hiring a tractorist) instead of supplying it from market, the economic benefits are not (only) technological (savings on production costs). The internal mode for governing of activity (specialization) often has substantial transaction costs advantages – economizing costs for finding best prices and suppliers, and for negotiating conditions of exchange; diminishing market uncertainty and risk from outside dependency, etc.

There are a great *range of specializations* of agricultural activities within internal organizations and/or across markets. In one extreme, a high level of outside transaction costs could restrict development of specialization in small-scale subsistence farming or caused multi-product integration into large member-oriented cooperatives. In another extreme, a specialized (in management) farm entrepreneur could carry out the entire production activity purchasing all related services (tilling, watering, fertilizing, plant protection, harvesting) from specialized markets[3]. Which mode for governing of a particular type of activity will dominate depends on *comparative* efficiency (advantages, disadvantages) of alternative forms.

The 'rational' agrarian agent would tend to choose the *most efficient mode(s)* for governing relations with others – that one which allow achieving maximum productivity (benefits) while minimizing *total* production *and* transaction costs[4]. According to the individual's *personal* characteristics (skills, ability) and the specific *institutional, market* and *natural* environment there will be *different* efficient forms for governing of farming activity. For instance, in transitional economies property rights were not well-defined and enforced, individuals had little managerial experience and face significant institutional, market, and behavioral uncertainty. In these conditions subsistence (semi-market) household holdings and large integrated cooperatives and agri-firms happened to me the most effective forms of farming organization (Bachev, 2004: 140). In the former case, the high transaction costs restricted agricultural specialization and farm size far bellow the technological opportunities (potential for growth in productivity). In the later case, the huge potential for transacting benefits extended enormously internal specialization and (horizontal and vertical) borders of farms beyond technological determinants.

[3] This is not a hypothetical but a typical for a good part of Japanese rice farming case.

[4] Often there is a trade-off between increasing productivity (through extension of specialization) and additional governing costs for coordination and exchange.

Following this new framework farm is to be studied as a *governance* structure with consumption, production, and transaction optimization functions (Bachev, 2004: 139). Furthermore, in order to explain evolution of diverse modes of specialization and distribution of activities between different farming organizations we have to analyze *structure of agrarian transactions* and associated transacting costs. Since much of the transaction costs are hardly to quantify the analysis is to concentrate on their *'critical factors'* – *institutional* (structure of formal and informal property rights and system of their enforcement), *behavioral* (agents bounded rationality, tendency for opportunism, risk aversion, reputation consideration), *dimensional* (frequency, uncertainty, assets specificity, and appropriability of transactions), and *technological*.

Next, in order to assess the *extend of specialization* of different farms we have to examine modes for governing of labor supply, service supply, inputs supply, land supply, finance supply, and marketing of output. For instance, intensified outside trade (inputs and service supply, marketing of farm produce) is an indicator for increased market specialization while on farm organization of additional labor (own production of inputs and 'services', internal utilization of outputs) is a sign for deepen internal specialization. What is more, 'full' efficiency of individual forms could be only understood in the context of governance of the *entire* farm. For example, effective extension of livestock farm could be done though new 'specialization' (e.g., production of forage for own animals though leasing in farmland and hiring crop labor) rather than pure enlargement of 'specialized' livestock operations (buying more forage from market, hiring additional livestock workers, selling more livestock products).

Finally, for evaluation of proper efficiency of diverse forms for agrarian organization the analysis is to embrace their *comparative* efficiency and *complementarities* as well as the larger *household* and *rural* economy. For instance, low productive multi-product cooperatives proved to be an effective form for governing of specialization and exchange in transitional countries with widespread small-scale and subsistence farming (Bachev, 2004: 141). On the other hand, a 'less' effective form of agricultural specialization such as part-time farming turned to be an essential part of households specialization (economy) in transitional conditions of high unemployment, great insecurity, and significant food costs.

III. Bulgarian Model for Farming Transformation

During the Communist period (1944-1989) Bulgarian farming was carried out in large public farms averaging thousands hectares and up to 6000 employees. Private ownership on major agrarian resources and market exchanges were abandoned. Public farms were responsible for food supply of local communities producing a broad range of farm and processed products. There were high functional and subject specialization of farms divisions and labor. Most activities (specialization and exchange) were governed by a Central Plan through an overcrowded multilevel hierarchical organization (reaching up to thirty per cent of agricultural employment). There were many reorganizations (Table 1) and experimentations with 'economic mechanisms' aiming at improving efficiency. Nevertheless, up to the collapse the socialist model the deficiency in incentives and productivity prevailed while most farms resources and production structure rested under state control. Since 1970s small 'private' farms were allowed mostly for food self-supply of households. Despite small size these 'personal plots' provided a major part of certain produce – maize, potatoes, eggs, meat, honey, tobacco, fruits and vegetable (National Statistical Institute).

Table 1: Development of Farming Structures in Bulgaria

Type of Farms		Communist Period					Transition		
	1944	1958	1969	1977	1985	1989	1995	2000	2005
Private farms (000)	1100	330	–	na	1600	1600	1772	755.3	515.3
Average size (ha)	3.9	2.6	–		0.4	0.4	1.3	0.9	1.8
Cooperatives	110	3200	795	–	–	–	2623	3125	1525
Average size (ha)	241	1264	4140				800	709.9	584
State farms	–	49	159	–	–	–	1002	232	–
Average size (ha)		3426	4040	–			338	358	
Agro-industrial complexes	–	–	–	143	298	–	–	–	–
Average size (ha)				32833	12600				
Collective farms	–	–	–		–	2101	–	–	–
Average size (ha)						2423			
Agro-firms	–	–	–		–	–	n.a.	2200	3704
Average size (ha)							n.a.	300	249.4

Source: National Statistical Institute.

Following 1991 reform all forcefully 'cooperated' or nationalized farmland was restituted to previous owners. Complex land transformation was implemented which took almost ten years to complete affecting eighty five per cent of agricultural land and turning a three-quarter of households into owners of farmland. Ancient cooperatives and other organizations established on their bases were liquidated and their assets distributed into individual shares. Liquidation took more than four years and made more than two million Bulgarians owners of small stakes in assets of ancient public farms.

Fundamental transformation of the economy was also carried out liberalizing markets, privatizing public enterprises, introducing EU institutions and standards. Economic reforms released market competition and introduced strong incentives for private entrepreneurship. More than 1,9 million private farms emerged on provisional or entirely restored private rights of lands and agrarian assets. By 1995 almost all agricultural activities (and specialization) were governed by entirely new market and private structures. Previous model of agricultural specialization within large public farms and a nationwide hierarchy was replaced by a capitalistic system of private entrepreneurship (private order) and free market mechanisms.

IV. Factors for Development of Farming Structures and Specialization

Specific type and pace of privatization of agrarian resources and big transitional uncertainty had important consequences for the development of farming organization in the country.

First, prolonged lack of full ownership on agrarian resources restricted feasible forms for their effective organization. Sells and long-term lease markets for farmland did not emerge until 2000 and annual lease was a major way for farm extension. Agrarian agents were unable to get full return on proprietary rights or use land ownership for setting up sustainable coalition and organization of other transactions (e.g. collateral against credit). On the other hand, unspecified or ideal character of ownership let rapid consolidation of fragmented farmland under management of small number of huge market-oriented enterprises (Table 2). Thus, governance of significant share of agrarian activities and specialization has been done within integral organizations rather than market competition.

Table 2: Share of Different Farms in Total Holdings and Major Resources in Bulgaria

Indicators	Physical persons	Coope-ratives	Sole traders	Com-panies	Associ-ations
Number of holdings with farmland (%)	99.0	0.3	0.4	0.2	0.05
Share in Utilized agricultural area (%)	30.3	40.3	11.7	16.1	1.6
Average size (ha)	1.4	592.6	118.8	352.5	126.2
Number of breeders without land (%)	96.1	0.2	1.9	1.7	0.1
Share in workforce (%)	95.5	1.2	0.8	1.4	0.3
Share in labor input (%)	91.1	4.1	1.4	2.8	0.6

Source: MAF, 2005.

Second, internal organization of available household resources in own individual or family farm was an effective way to overcome a great institutional and economic uncertainty and minimize transacting costs. During much of the transition market and contract trade of household's capital was either impossible or very expensive – missing markets situation, high uncertainty and risk, asymmetry of information, big possibility for opportunism in time of hardship. Many lost their jobs as result of restructuring of public companies. Low-payoff of outside trade was combined with an increased share of households' food costs. Internal organization turned to be most-effective way to protect and get return on resources, and secure stable income for households. A long-term tradition with 'personal plots' from the Communist period, and insignificant costs for acquiring specialized knowledge made development costs for own farm accessible for everybody regardless previous occupation. Own production has been effective mode to guarantee cheap, stable and high-quality delivery of food, provide employment for family members, or be favorable free-time occupation.

That is how a huge subsistent and part-time farming emerged affecting a considerable part of households. Miniature (full time) specialization of unemployed/retired persons or (part-time) diversification into farming of engages in other businesses took place. Even now full-time employed in the sector accounts for eleven and a half per cent of all workers in the country (National Statistical Institute, 2005). In addition, almost one million Bulgarians are involved in farming on a part-time base as 'supplementary' income source. Estimates in Annual Work Units (AWU) show that agriculture comprises more than 26 per cent of the overall employment (MAF, 2006). Labor contributed by part-time workers reaches fifty three per cent of AWU of the sector.

Now more than three-quarters of farms are less than one ha averaging a half ha. More than ninety seven per cent of livestock holdings are unprofessional farms having few heads but breading ninety six per cent of goats, eighty six per cent of sheep, seventy eight per cent of cattle, and sixty per cent of pigs. More than ninety per cent of farms are less that two European Size Units[5], and generally considered as subsistent and semi-subsistence farms. In livestock grazing and mix orientation later bring considerable share of Standard Gross Margin in respective groups (MAF, 2005).

Within small farms no labor specialization is practiced or it is carried out between few family members. Most farms are not specialized for market rather aim at serving diversified food needs of households producing a great range of crop and livestock products. For vegetables, fruits, vine and livestock a significant portion of the overall national output is for own consumption (MAF, 2005). A great fraction of farms sells out only surpluses for major commodity products (Figure 1). For Physical Persons the number is higher as less than thirty nine per cent of farms report they sell products and for more than fifty per cent those are surpluses not consumed by households (MAF, 2005).

Figure 1: Share of Farms Marketing Products in Bulgaria

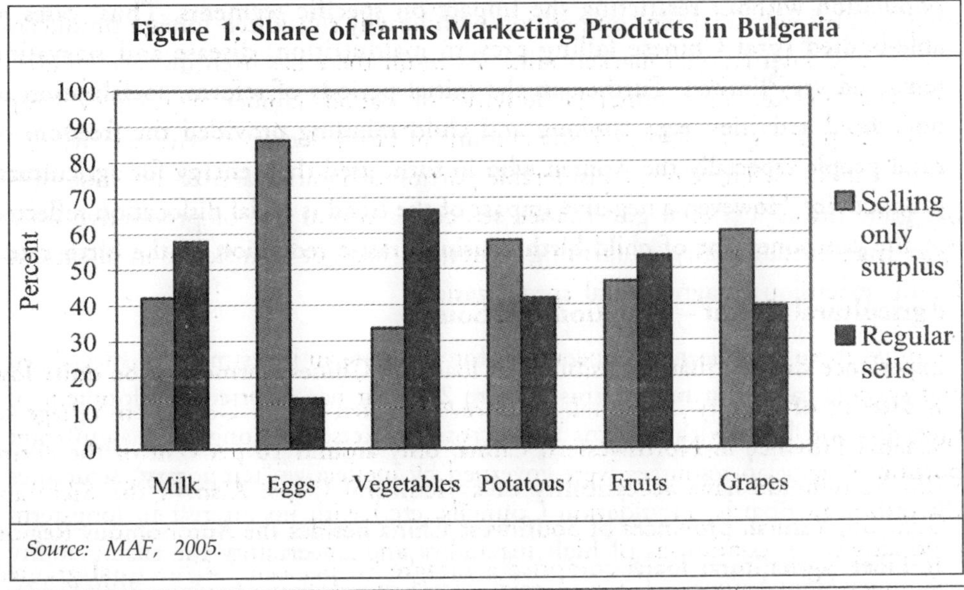

Source: MAF, 2005.

[5] One European Size Unit = 1200 Euro.

Third, most agrarian agents found their skills (previous specialization, team working tradition) and portions in acquired agrarian resources in a high interdependency. A great part of individuals' shares in ancient public farms were in indivisible assets – large machinery, buildings, irrigation facilities. Later there was no alternative but liquidate (through sales, consumption) or keep them up as a joint (cooperative) ownership. In many cases, 'new' land owners restituted their plots with fruit trees, vineyards, etc., and they could practically execute much of activities in cooperation. Most land and share-holders happened to live away from rural areas, or have other business, or be old of age, or have no skills or capital to start own farms. In absence of big demand for farmland and confidence in emerging new private modes, the only option was to joint a cooperative. That is how more than 40 per cent of new owners pulled their free land, assets, and labor in new production cooperatives. Similarly, most privatized state farming and livestock companies were taken-over by 'managerial-workers' teams and registered as Shareholder Companies and Associations keeping previous (internal and market) specialization.

Fourth, there was no or little experience in managing private or collective farms in majority of new entrepreneurs. Moreover, there has been lack of public support in farming training, advisory service and funding (Bachev, 2006: 143). That has been coupled with a strong competition with heavily subsidized foreign producers at local and international markets alike. As result there has been massive failures, take-overs and ceasing commercial activities of a good number of newly evolving farms. Comparing to 1994 pick the amount of farms decreased with seventy two per cent in 2005. Therefore, considerable portion of agrarian activity and resources has gone through several governance structures or got out of productive uses. That little sustainability of farming structures has got significant implications for the extension of agricultural specialization.

Fifth, there have been low incentives for a long-term investment in specialized and specific capital in most farms (Figure 2). That has deterred development of specialization both within farms and across markets. For long time significant portion of agrarian resources were governed by ineffective "temporary" structures (Privatization Boards, Liquidation Councils, etc.) with no interest in long-term productivity. In conditions of high instability and uncertainty the sustainability and investment activity of commercial enterprises is low (Bachev, 2006:137).

Moreover, much of farming-related investment is highly specific and can hardly be funded by outside credit or equity sells (Bachev, 2006: 137). Majority of small-scale commercial farms are run by older generation entrepreneurs with a short business (investment) perspective[6]. Almost all farmland of large business farms has been supplied by provisional lease-in contracts. While there have been strong investments in mobile material assets there has been no long-term investment for improving land productivity (renovation of orchard, vineyard, irrigation facilities; compensating N, P and K intakes). Most cooperatives has been mismanaged or experienced significant funding problems because of different investment preferences of members (Bachev, 2006: 142). As far as subsistence farming is concerned it does not necessitate significant investment since households demand is stable.

V. Impact of Farming Transformation on Agricultural Specialization

Post-communist transformation of agriculture has increased farming importance in national economy both in terms of contribution to the Gross Domestic Product and employment (Figure 2). Comparing to 1989 agricultural employment increased by twenty per cent as its share in overall employment doubled. Therefore, overall agricultural specialization expanded as more people find work and income in this sector of national economy. On the other hand, seventy five per cent of employed in farming are engaged on part-time bases indicating domination of

Figure 2: Share of Bulgarian Agriculture in National Economy

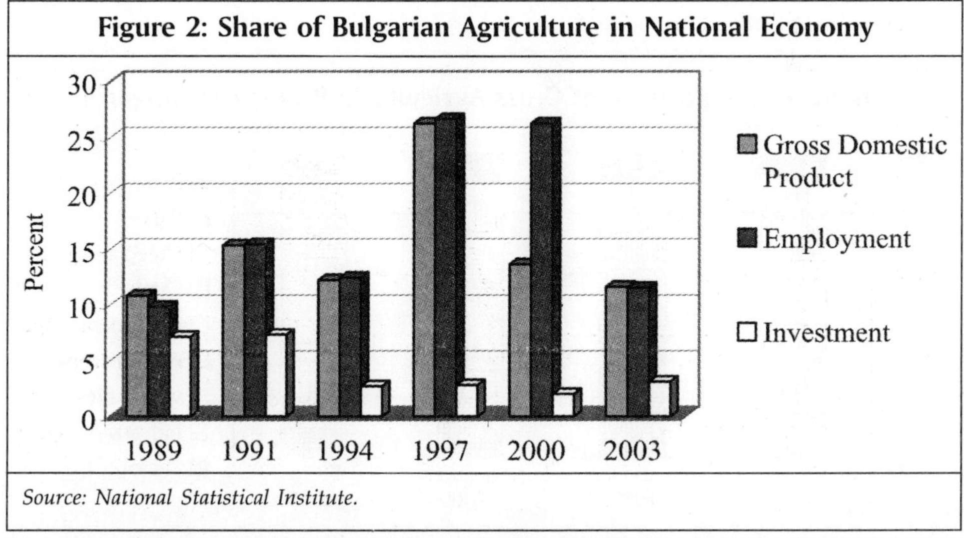

Source: National Statistical Institute.

6 Managers older than forty five and sixty five are eighty five per cent and forty per cent accordingly (MAF, 2005).

primitive (rather that complete) forms of agricultural specialization (less specialization of labor, less exchanges).

Post-communist market adjustment and farm adaptation has been associated with alteration of production structure and a general decline in agricultural output (Figure 3). There has been a significant change in general product specialization and importance of different sub-sectors (Figure 4). Livestock lost its dominant

Figure 3: Dynamics of Major Agricultural Productions (1989=100)

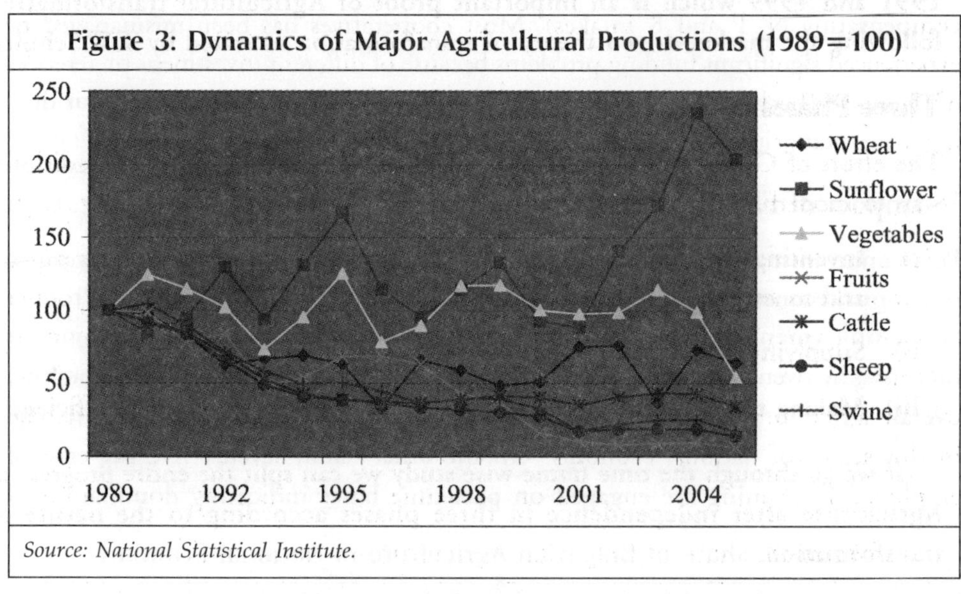

Source: National Statistical Institute.

Figure 4: Composition of Gross Agricultural Product in Bulgaria

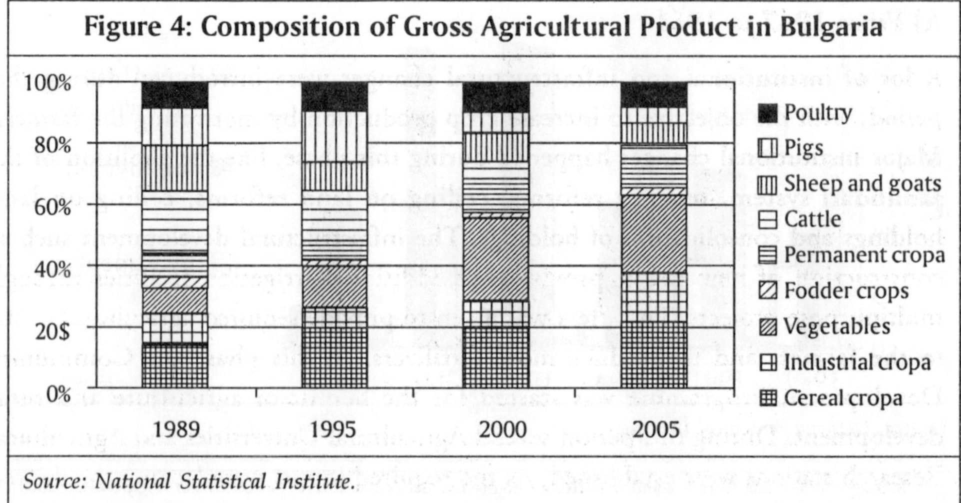

Source: National Statistical Institute.

pre-reform share melting down to twenty six per cent of the Gross Agricultural Product (GAP). While poultry maintains its segment, cattle, sheep and goat, and pigs portions dropped considerably. At the same time, vegetables have seen its share significantly augmented expanding from the sixth to the largest contributor to GAP. Cereals and industrial crops also extended being the second and the third most important productions.

Newly evolving farming structure is characterized with entirely different type of specialization comparing to the pre-reform period. A sizable share of farms produce large number of different products. Only in cereals, oil plants, aromatic and medicinal crops production is more specialized. A fraction of specialized livestock farms (big operators) is insignificant. While different 'mix farms' and primitive 'grazing animal farms' comprise a bulk of the farms, specialized farms of various kinds produce the greatest part of Standard Gross Margin (SGM) of the sector (Figure 5). Larger commercial farms (bigger than two ESU) with different type of specialization contribute most to SGM in all groups of specialization. A tiny number of all farms in permanent crops (zero point three per cent), field crops (one point six per cent), pigs and poultry (zero point two per cent), horticulture (zero point forth per cent), and mix cropping (zero point twenty five per

Figure 5: Share of Farms with Different Specialization and Contribution to Standard Gross Margin of Agriculture

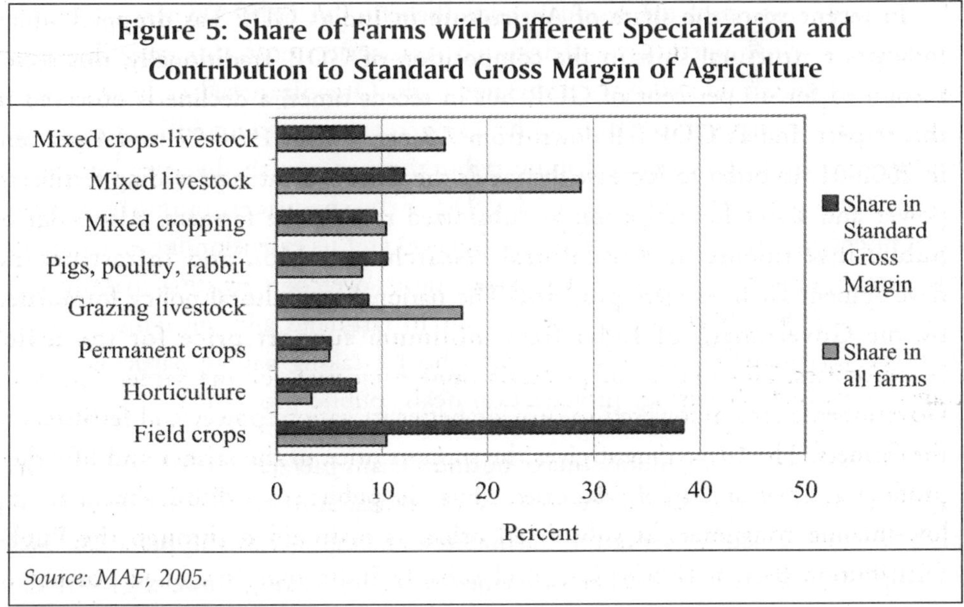

Source: MAF, 2005.

cent) are bigger than hundred ESU but they produce a significant share of SGM in these groups – sixty three per cent, fifty four per cent, fifty four per cent, twenty six per cent, and twenty eight per cent accordingly (MAF, 2005).

VI. Governing and Specialization of Business Farms

Our survey has found out that agro-firms are commonly large specialized enterprises. Most of these firms were set up as family/partnership businesses during first years of transition by younger generation entrepreneurs. Specific management skills and social status, and combination of partnership assets (technological knowledge, business ties, available resources) led to a rapid extension of farms through enormous concentration of management/ownership of resources, specialization and exploration of economy of scale/scope, and modernization of enterprises (Bachev, 2006: 136). Institutional uncertainty, unsettled rights on assets, personal relations and 'quasi'/entirely integrated modes were extensively used to overcome transaction difficulties. Number of agro-firms has increased 20 folds since 1990 and doubled since 2000 as their share in overall resources augmented (National Statistical Institute).

Business farms are profit-oriented organizations, and farmer(s) have great incentives to invest in farm-specific (human, material, intangible) capital because they are sole owners of the residual rights (benefits) of farm. Owners are family members or close partners, and internal transaction costs for coordination, decision-making, and motivation are not high. Organizational style of *firm* is preferred since it provides opportunity to overcome coalition difficulties (joint ventures with outside capital, dispute rights through court); diversify into farm related/independent businesses; develop firm-specific intangible capital (advertisement, brand names, public confidence) and its extension into a daughter company, trade and transfer through generations; overcome existing institutional restrictions (e.g., direct foreign investments in farmland and engaging in trade with cereals/vine/dairy); provide explicit rights for taking part in particular types transactions (e.g., licensing, privatization deals, public programs).

The large size and reputation make business farms preferable partners in inputs supply and marketing deals. Recurrence of transactions with same partners is high which restricts information asymmetry and opportunistic behavior, and develops mutual trust and other mechanisms for facilitating (lowering costs of) relationships – planning, adjustment and payment modes, guarantee schemes, dispute resolution devices. Agro-firms have giant negotiating power and effective economic, political, etc. mechanisms to enforce contracts. They possess a great potential to collect market information, search for best partners, use experts and innovation, meet special (collateral) requirements and bear risk and costs of failures. They could explore economy of scale/scope on production and management ('package' arrangement of credits for many projects and interlinking inputs supply with know-how supply/crediting/marketing). They are also able to invest considerable relation-specific capital (information, expertise, reputation, lobbying) for dealing with funding institutions, agrarian bureaucracy, and market agents at national or international scale.

Under conditions of non-working court and contract enforcement systems, all critical farm transactions are governed (controlled/protected) through internal modes. Farm-specific assets such as critical machinery, vineyards, orchards, animals, processing facilities, and adjoining land, are all safeguarded by ownership mode. Low cost standard (one-season, share rent) lease-in contracts are widely used to govern land supply from tens/hundreds of proprietors.

'Critical' transactions are integrated through extensive labor employment (Figure 6). Preference to 'own making' (internal specialization and organization of activity) rather then market procurement (buying specialized inputs and services across market) is determined by transaction costs economizing reason – the high farm specificity and/or uncertainty associated with certain transactions. Besides,

Figure 6: Governing of Labor Supply in Bulgarian Farms

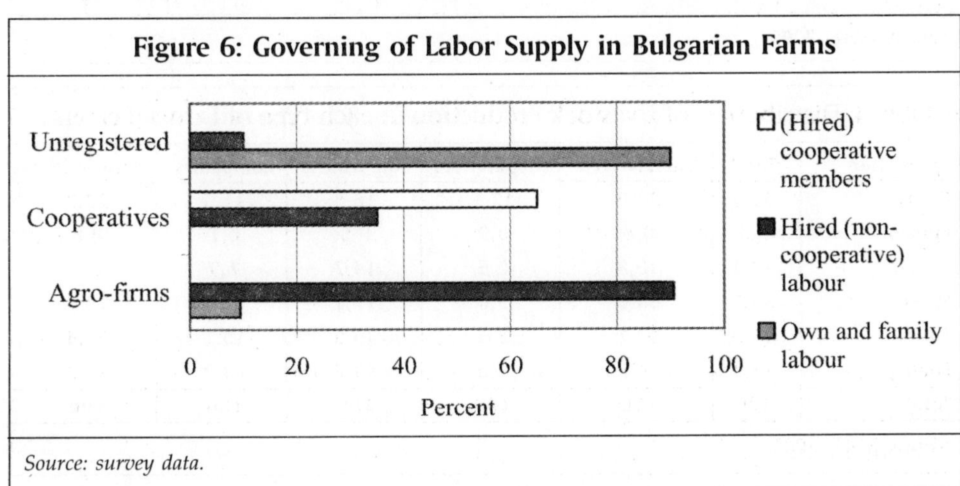

Source: survey data.

core labor (specialists, mechanists) is hired on a permanent basis and special forms such as output-based compensation, interlinking (housing, services), social disbursements, paid holidays are further used to enhance motivation. Furthermore, in large business farms it is typical strong internal division and specialization of labor – both functional (management, production, marketing, security) and subject (crop, livestock, services). Nevertheless, in farming hired (non-family) labor command high transacting costs for directing, supervising, protecting opportunistic behavior, disputing. Therefore large business farms are mostly specialized in less labor-consuming, highly-intensive and standardized productions (cereals, sunflower, aromatic and medicinal plants, poultry, and pigs) where control of labor is easy or output-based compensation could be effectively applied (Table 3 and 4). These farms are also a major provider of mechanization services to other farms.

Table 3: Distribution of Crop Production in each type of Farms (Percent)

Production	Unregistered		Cooperatives		Sole traders		Companies		Associations	
	Farms	Area	Farms	Area	Farms	Area	Farms	Area	Farms	Area
Cereals	51,1	59,5	93,0	59,7	54,3	62,3	55,1	61,1	51,1	56,0
Dry pulses	13,0	0,8	5,7	0,6	5,9	0,6	3,6	0,5	5,9	1,1
Oil crops	5,1	17,8	74,4	28,6	28,9	29,7	41,6	27,0	28,2	26,0
Textile crops	0,0	0,0	1,0	0,2	0,1	0,0	0,4	0,2	0,5	0,2
Aromatic and medicinal plants	0,5	1,7	17,6	2,5	4,8	1,9	12,8	4,4	6,7	5,6
Forage	28,6	7,5	29,6	3,1	21,2	1,2	14,0	1,7	23,4	7,5
Vegetables	49,2	5,5	8,3	0,2	29,2	0,5	11,9	0,4	21,2	0,4
Fruits	8,7	2,0	14,7	0,9	11,4	0,3	12,2	0,7	35,2	9,5
Total	100	100	100	100	100	100	100	100	100	100

Source: MAF, 2005.

Table 4: Distribution of Livestock Production in each type of Farms (Percent)

Production	Total	Unregistered	Cooperatives	Sole traders	Companies	Associations
Cattle	31,8	32,0	13,1	18,8	11,3	22,8
Buffalo	0,4	0,4	0,2	0,5	0,7	0,8
Goats	40,4	40,8	0,5	14,7	2,9	6,5
Sheep	35,7	36,0	5,0	17,1	6,3	13,7
Pigs	41,9	42,1	3,6	29,9	15,1	20,4
Poultry	74,3	74,8	1,2	43,2	13,2	15,3
Total	100	100	100	100	100	100

Source: MAF, 2005.

Principally, own supply ('making') rather than outside procurement ('buying') is common for essential services and inputs (Table 5 and 6) which prevents risk from unilateral dependency (opportunism of supplier) or missing market situation. In the case of high asset interdependency with downstream partners (product specificity; quality/quantity dependency) reciprocal supply of 'inputs against marketing' is applied. Our survey has found out that funding is secured through an effective combination of equity, debt, public and hybrid modes. Standard activities/assets are financed by bank credit since it is easy to arrange a loan. Alternatively, farm-specific investments are financed through private modes – own sources, 'personal' loans and co-investment. Special contract modes are used to mitigate funding difficulties (shortage of working capital) or facilitate mutually-dependent relations with buyers/suppliers such as delayed payments for inputs supply (zero interest, loans in kind), interlinking credit with inputs supply and marketing, leasing or accepting outside investment ('hostage taking', joint ownership) of long-term assets. Agro-firms are also quite successful in recently introduced public support programs (e.g., EU Special Assistance Programme for Agricultural and Rural Development) developing good proposals, meeting formal requirements, dealing with complicated paper work, and 'arranging' selection of projects.

Table 5: Governing of Service Supply in Bulgarian Farms (Percent of Farms)

Service type	Modes	Unregistered	Cooperatives	Agro-firms
Technological knowledge and advises	Own supply	24	49	65
	Own cooperative	5	7	15
	Market supplier	13	10	25
Mechanization services	Own supply	18	85	60
	Own cooperative	22	0	18
	Market supplier	15	15	28
Spreading chemicals and pesticides	Own supply	40	65	60
	Own cooperative	15	7	12
	Market supplier	12	25	28
Veterinary services	Own supply	20	60	40
	Own cooperative	5	0	0
	Market supplier	40	40	60

Source: survey data.

Table 6: Governing of Inputs Supply in Bulgarian Farms (Percent of Farms)

Inputs type	Supplier	Unregistered	Cooperatives	Agro-firms
Chemicals	Own production	17	0	0
	Own cooperative	10	5	15
	Market supplier	55	95	90
	Buyer of farm output	24	13	33
Seeds and seedlings (crop farms)	Own production	47	53	33
	Own cooperative	3	15	23
	Market supplier	50	32	45
	Buyer of farm output	4	41	44
Forage (livestock farms)	Own production	55	65	50
	Own cooperative	0	0	35
	Market supplier	45	35	15
	Buyer of farm output	9	6	53
Machinery	Own production	12	13	0
	Own cooperative	20	17	46
	Market supplier	68	70	54
	Buyer of farm output	15	0	19
Livestock	Own production	37	50	28
	Own cooperative	21	31	33
	Market supplier	42	19	39
	Buyer of farm output	40	17	13

Source: survey data.

In marketing of farm output and services a classical trade across market (wholesale market; business with market agents) dominates (Table 7). Main part of farm's product has standardized (commodity) character and market prices/ competition effectively governs relations with partners. When specificity of output to a particular buyer (processor, retailer) is high (technology, quality, time of delivery, site-specificity) then delivery contracts with the respective partner are employed to tailor or protect transactions. Here division of labor and specialization in vertical chain is governed by a private order rather than classical market competition. Intra-firm processing and retailing is also practiced by some farms. That further extends farm boundaries through internal organization and specialization of interdependent activities. Larger operational size and frequency of transacting provide an economic opportunity for internal exploration of

Table 7: Governing of Marketing in Bulgarian Farms (Percent of Farms)

Output	Modes	Unregistered	Cooperatives	Agro-firms
Grain	Own cooperative	9	7	9
	Another farm/firm	50	85	75
	Processor	25	39	37
	Retail	6	7	16
Vegetables	Own processing	0	0	15
	Another farm/firm	24	24	35
	Wholesale market	6	5	15
	Processor	38	66	30
	Retail	12	0	6
Fruits and grape	Own processing	15	7	19
	Own cooperative	24	7	9
	Another farm/firm	48	39	32
	Wholesale market	0	22	22
	Processor	15	36	25
	Retail	6	0	0
Meat	Own processing	0	10	15
	Another farm/firm	65	71	80
	Processor	29	43	30
	Retail	15	36	20
Milk	Own processing	0	10	15
	Another farm/firm	42	43	40
	Processor	51	64	45
	Retail	19	0	15

Source: survey data.

interdependent assets (farming-processing-retailing). Vertical integration helps protect dependent investments and payoffs from marketing of processed products – getting full profit, brand name trade, lessened market dependency (easy storage, transportation).

VII. Governing and Specialization of Production Cooperatives

Cooperatives are the biggest farms in terms of land and labor management (Table 2). Coops concentrate a major part of cereals, oil and forage crops, orchards and vineyards, and they are a key service providers for their members and rural population.

The cooperative was the single most effective form of organization in the absence of settled rights for main agrarian resources and/or inherited high interdependence of available assets (restituted farmland, acquired individual shares in the actives of old cooperatives, narrow specialization of labor) (Bachev, 2006: 141). Moreover, most cooperatives developed along with small-scale and subsistent farming. Namely, 'not-for-profit' character and strong membership (rather than market) orientation attracted many households. As for production, the coop was perceived as effective (cheap, stable) form of supplying highly specific to individual farm inputs and services (feed for animals; mechanization; storage, processing, and marketing of output) as well as food for households. The cooperative, rather than other formal collective (e.g., firm) forms, has been mostly preferred. Coops were initiated by older generation entrepreneurs and tradition has played a role. Besides, this mode allows individuals an easy (low-cost) entrance and exit keeping control over a major resource (land) and permitting a democratic participation in and supervision over management. In addition, cooperative form provides some important tax advantages (exemption from sale transactions with members, and received rent in kind) and possibilities for organizing transactions that are not legitimate for other modes (e.g., credit supply, marketing, and lobbying nation-wide).

Larger operational size gives cooperatives a great opportunity for efficient use of labor (teamwork, division and specialization of work), farmland (cultivation in big consolidated plots, effective crop rotation), and material assets (exploration of economy of scale/scope of large machinery). In addition, they have a superior potential to minimize market uncertainty ('risk pooling', advertisement, storing, integration into processing and marketing), and organize critical transactions (access to credit; negotiating positions in input supply and marketing; facilitate land consolidation through lease-in and lease-out deals; technological innovations), and invest in intangible capital (reputation, labels, brand names).

Cooperative activities are not difficult to manage since internal (members) demand for output and services is known and 'marketing' secured. In addition, coops concentrate on few highly standardized (mass) products with stable market and profitability. All these assists financing: advance funding of activities commissioned by members is commonly practiced, while production of universal commodities is easily financed by public programs or commercial credit.

Furthermore, coops offer low-cost long-term leasing of land. That is often coupled with simultaneous lease-out deals as a specific mode for cashing coops output or facilitating relations between landowners and private farms. The integral organization of critical 'services' and inputs supply is broadly practiced. Output-based payment of labor is common, which restricts opportunism and minimizes internal transaction costs. Besides, cooperatives provide employment for members who otherwise would have no other job opportunities – housewives, pre- and/or retired persons. They are preferred employers since they offer higher job security, social payments, paid holidays, etc. Marketing risk is governed by effective delivery contracts or integrated into own processing. In a situation of 'missing markets' in rural areas, the cooperative mode is also the single form for organizing important activities such as bakeries, wholesale, retail trade, etc. Given considerable transacting benefits, most coop members accept lower than market returns on their resources – lower wages, inferior or no rent for land and dividends for shares.

There have been some adjustments in size of coops, memberships, and production structure. Number of them have moved toward corporate ('new generation') type of governance, applying profit-making goals, closed-membership policies and joint-ventures with other organizations. At the same time, cooperatives show certain disadvantages as form for farm organization. Large coalition makes individual and collective control over management very difficult (costly) providing possibilities for mismanagement (on-the-job consumption, unprofitable members' deals). Besides, there are differences in investment preferences of the diverse members (old-younger; working-non-working; large-small shareholders) due to the non-tradable character of cooperative shares ('horizon problem'). Given the fact that most members are old in age, small shareholders, and non-permanent employees, the incentives for long-term investment in cooperatives have been very low. Finally, many coops fall short adapting to diversified (service) needs of members and exploring potential of inter-cooperative modes. Accordingly, cooperatives' long-term efficiency diminishes considerably in relation to market, contract and partnership modes, and almost forty per cent of coops have gone bankrupt or ceased to exist in last five years.

VIII. Governing and Specialization of Unregistered Farms

Majority of commercial farms are 'unregistered farms'[7] which are mainly in labor-intensive productions (vegetables, tobacco, vineyards, berries, melons, flowers, mushrooms, medicinal and aromatic crops, livestock, sericulture, bee kipping) and natural meadows. Those are predominately individual or family holdings, and farm size is exclusively determined by available household resources – farmland, labor, finance. Internal governing costs are insignificant since transactions are between family members (common goals, high confidence, and no cheating behavior dominates) or not existing at all (one-person farm). Small collective organization is also practiced for some production activities which allows a partial exploration of economies of scale or make a part-time farming practically possible. The former mode is cost-effective since transactions are not complicated and easily controlled. Group members are usually close friends, neighbors, or relatives, and mutual trust and self-restriction of opportunism govern relations.

Commercial farmers have strong incentives to adapt to market demand and increase productivity (intensifying work, investing in human and material assets) since they own the whole residuals (income). Extension of farm size through outside supply of labor or services is restricted since directing, monitoring, and disputing costs are extremely high in labor-demanding and spatially dispersed productions. External financing of farming via debt, equity sell, or preferential public programs have been out of reach because of the high costs for preparing project proposals; meeting formal paperwork, ownership, coo-financing, etc., requirements; and 'arranging' funding. Thus possibility for effective farm enlargement and growth in productivity through intra-farm division and specialization of labor, mechanization, application of chemicals, innovation has been limited by the small internal investment capacity (savings, profit). In general, primitive technologies and poor environmental and animal welfare standards prevail (Bachev, 2006: 143). As much as forty per cent of surveyed farms report not using essential services at all. Low-cost, outside land supply (leasing) is practiced by commercial farms to explore economies of scale on existing assets. Outside supply of indispensable inputs and services (seeds, chemicals, veterinary) is not

---

7 In Bulgaria there is not a formal requirement to register a farm. Official statistics report as 'Physical Persons' unregistered farms meeting certain criteria (minimum size of farmland, number of animals, etc.).

connected with significant costs since they have occasional and standardized character (low-specificity, many suppliers). In contrast, highly-specific feed supplies for animals and mechanization services are effectively secured through joint ownership modes such as cooperative and group farming.

Own farm enterprise has been secure mode for providing (full or part-time) employment for family members. Family organization is also an effective form for intergeneration transfer of farm-specific intangible assets (know-how, reputation). However, small-scale commercial farms have little ability to meet institutional and market restrictions, and bear the risk, and protect against natural and market hazards. A great number of them face great transacting difficulties in marketing of their output. Most often they are not preferable partners for big buyers because of small volume and less-standardized character of output, and impossibility (unaffordable costs) to verify quality of products through laboratory tests, certificates, etc. On the other hand, official wholesale markets have been inaccessible for these farms for the reason of great distance; high fees; and requirements for volume, special preparation, certification. Besides, farms frequently experience no accomplishment of contract obligations (none or delayed payment), huge market price fluctuation, monopoly situation, missing markets etc. In order to protect transacting and avoid unwanted exchanges primitive forms for risk minimization is commonly used such as investment in more universal but less productive (profitable) assets, diversification of production, informal cash and carry deals, direct retail marketing, etc.

With exception of tobacco producers, the development of effective collective organization for risk sharing, price negotiation, marketing, and/or lobbying for public support, have been difficult. That has been because of the high transacting costs ('free riding' problem), diversified interests of individual farmers (old-young farmers; larger or smaller-size farms; specialized-diversified operators, etc.), and low reputation and inefficiency of new emerging farmers associations. Majority of small commercial farms are vulnerable and have poor mechanisms to protect from outside institutional, market, and natural disturbances. Most of them have little ability to face severe market competition, and meet fast evolving institutional (e.g., EU) restrictions, and bear the risk, and safeguard against natural and market hazard (buy insurance, diversify, or cooperate). All these bring about to a significant income variation for individual farms, sectors, regions, and different years. What is more, farms in entire sub-sectors (like dairy) have been unable to adapt to new market and institutional order associated with EU integration. Consequently, there has been a constant process of transfer of land management toward bigger farms, restriction/ceasing of commercial activities, and decreasing the number of small farms.

Last but not least important, unlike other forms of organization the life cycle of one-person (family) farm is greatly determined by the age of the entrepreneur. Besides, incentives for a long-term investment in specialized assets for increasing sustainability are low for older farmers since there is no secondary market for farm-specific assets (investments in human capital, good reputation, know-how, organizational modernization). Therefore, a good number of small commercial farms operate at low sustainable level given the fact that most of the farm managers and labor are old in age.

IX. Conclusion

Post-communist transformation of Bulgarian agriculture let us determine factors, modes, and extend of farming specialization and governance in a fast changing institutional and market environment. We have demonstrated that particular type and pace of institutional modernization and market adjustment are responsible for evolution of new and specific (quite different from other European countries) forms for farming organization and specialization. Bulgarian model is characterized with a huge subsistence and part-time farming, production cooperation at a great scale, enormous concentration of resources in few business enterprises, widespread use of vertically integrated modes, domination of informal modes and personal relations, numerous missing markets and failures, backward technological 'development', significant government failures.

The high efficiency and sustainability of all these structures could be only explained by the specific institutional restrictions and comparative transaction costs and benefits of individual modes. The later eventually determines the kind and extend of agricultural specialization within the farm (internal mode) and across markets (between farms). In one case, it leads to a great restriction of the farm size and specialization far below the technologically optimal level (subsistence, part-time and small-scale farming). In other instances, the same factors are responsible for enormous integration of activities and extension of farms boundaries beyond the technological determinants.

(Hrabrin Bachev, Institute of Agricultural Economics, Sofia, Bulgaria. The author can be reached at hbachev@yahoo.com).

References

Bachev H. (2004): "Efficiency of Agrarian Organizations", in: *Farm Management and Rural Planning 5*, Kyushu University, Fukuoka, 135-150.

Bachev H. (2006): "Governing of Bulgarian Farms – Modes, Efficiency, Impact of EU Accession", in *Agriculture in Face of Changing Markets, Institutions and Policies: Challenges and Strategies*, ed. Curtiss, Balmann, Dautzenberg and Happe, IAMO, Halle, 133-149.

Bachev H. and M.Tsuji (2001): "Structures for Organization of Transactions in Bulgarian Agriculture", *Journal of the Faculty of Agriculture of Kyushu University*, No 46 (1), Fukuoka 123-151.

Bencheva N. (2005): "Transition of Bulgarian Agriculture – Situation, Problems and Perspectives for Development", *Journal Central European Agriculture*, Vol.6, No4, pp.473-480.

Coase R. (1937): "The Nature of the Firm", *Economica* 4, 386-495.

Furuboth E. and R. Richter (1998): *Institutions and Economic Theory: The Contribution of the New Institutional Economics*, Ann Arbor, The University of Michigan Press.

Masten, S.E. (1991): "Transaction-cost economics and the organization of agricultural transactions", Chicago, IL, presented at the NC-194 World Food Systems Project Symposium, pp.17-18.

MAF (2005): "Agricultural Census in Bulgaria 2003 Results", Ministry of Agriculture and Forestry, Sofia.

North D. (1990): *Institutions, Institutional Change and Economic Performance*, Cambridge University Press.

OECD (2000): "Review of Agricultural Policies in Bulgaria", OECD, Paris and Sofia.

Sporleder T. (1992): "Managerial Economics of Vertically Coordinated Agricultural Firms, American *Journal of Agricultural Economics*", Vol. 74, No. 5, pp. 1226-1231

Williamson O. (1996): "The Mechanisms of Governance", *New York*, Oxford University Press.

8

Agricultural Extension in Africa and Asia*

Carl K Eicher

The current paper attempts to review the role of agricultural extension systems in helping smallholder farmers in Africa and Asia increase agriculture production and their livelihoods. The paper tries to explain the reasons why agricultural extension is an important framework to incorporate into the information and knowledge systems for small farmers. The paper has highlighted certain extension models in various stages of development and implementation in developing world like the National Public Extension Model, the Commodity Extension and Research Model, the Training and Visit (T&V) Extension Model, the NGO Extension Model, the private extension model and the Farmer Field School (FFS). The paper in particular focuses on the T&V extension model and the growth of FFS model along with the extension of the ATMA (The Agricultural Technology Management Agency Model).

* Literature review prepared for the World AgInfo Project, Cornell University, Ithaca, New York.

Source: World Ag Info Project, August 15, 2007 (http://www.worldaginfo.org).

The Gene Revolution requires a much higher level of institutional capacity than the Green Revolution in terms of agricultural research, bio-safety regulation, intellectual property protection, and agricultural input and output markets.

– Prabhu Pingali, 2007.

Background

Without question, agricultural extension is now back on the development agenda. The acknowledged failure of the T&V extension model in Asia and Africa in the late eighties and early ninties has stimulated debate on extension reforms and new extension models such as Farmer Field Schools. Today, extension reforms are underway in many countries in Asia and Latin America and to a lesser extent in Africa. The purpose of this review is to summarize the literature on the role of agricultural extension systems in helping smallholder farmers in Africa and Asia increase agriculture production and their livelihoods.

Five reasons explain why agricultural extension is an important institution to incorporate into the design of information and knowledge systems for smallholders:

- First, extension departments form the centerpiece in most Ministries of Agriculture in terms of the number of personnel. For example, there are an estimated 100,000 extension agents in India and 10,000 in Ethiopia and the government of Ethiopia is training another 33,000.
- Second, because of calls for market-driven development, the architects of the dominant public extension, research and agricultural education models in both Africa and Asia are being challenged to look for less costly and more pluralistic systems that can be privatized or served by nongovernmental organizations (NGOs) (Antholt 1998 and Anderson 2007).
- Third, is a growing recognition that extension must go beyond transferring new food crop technology to farmers and focus on helping the rural poor by promoting agricultural diversification, increasing rural employment, and helping farmers gain access to biotechnology (Pingal 2007) and access to export markets (Van den Ban and Samantha 2006).
- Fourth, the rapid spread of the Farmer Field School (FFS) extension model has encouraged donors to put extension back on the development agenda.

The FFS model is a community learning model that has been adopted in some 50 to 70 countries over the past 15 years. However, numerous economists argue that the model is not financially sustainable after foreign aid is withdrawn. (Feder *et al.,* 2004a).

- Fifth, the private sector is playing an important role in extending information to smallholders growing GM crops in India, China, Brazil and South Africa. This new development calls for research on private and new forms of private-public extension models (Anderson and Crowder 2000).[1]

Conceptual Issues

The diffusion of technology has been a powerful source of economic change for generations. During the 1940s and 1950s, diffusion research emerged in rural sociology departments in the United States (Ruttan 2003). By the 1960s, these traditions were continued in communications, geography, marketing and economics. For a discussion of diffusion research, see Rodger's now classic *Diffusion of Innovations,* first published in 1962 and updated in 1971, 1983, 1995 and 2003 and the important work by Roling (1988 and 2006).

Before we discuss the basic extension models, it is important to note how extension is linked to development institutions such as producer groups, research institutions, agricultural higher education and input and product markets. The paradigm of agricultural knowledge and information system (AKIS) which flourished in the 1990s stressed the importance of developing a *system* of institutions (e.g., research, extension and education) that cooperate and communicate with each other to achieve an overall goal of increasing agricultural productivity (Figure 1).

However, the AKIS paradigm has been criticized because of its linear vision on delivering technology to large farms while ignoring farmers with limited land and resources, and not listening to farmers in terms of their problems and priorities for government and university researchers. Therefore, the conceptual debate has now shifted from the AKIS paradigm to agricultural innovation systems (Sulaiman

---

1 For example, management may become the responsibility of farmer or agribusiness organizations rather than local governments. Extension can still be publicly funded, but funds can flow through farmer organizations that have a controlling interest in fund allocation. Farmer organizations, in turn, may contract out extension services to private providers and NGOs as in Uganda's National Agricultural Advisory Services (World Bank forthcoming).

Figure 1: Agricultural Knowledge and Information Systems

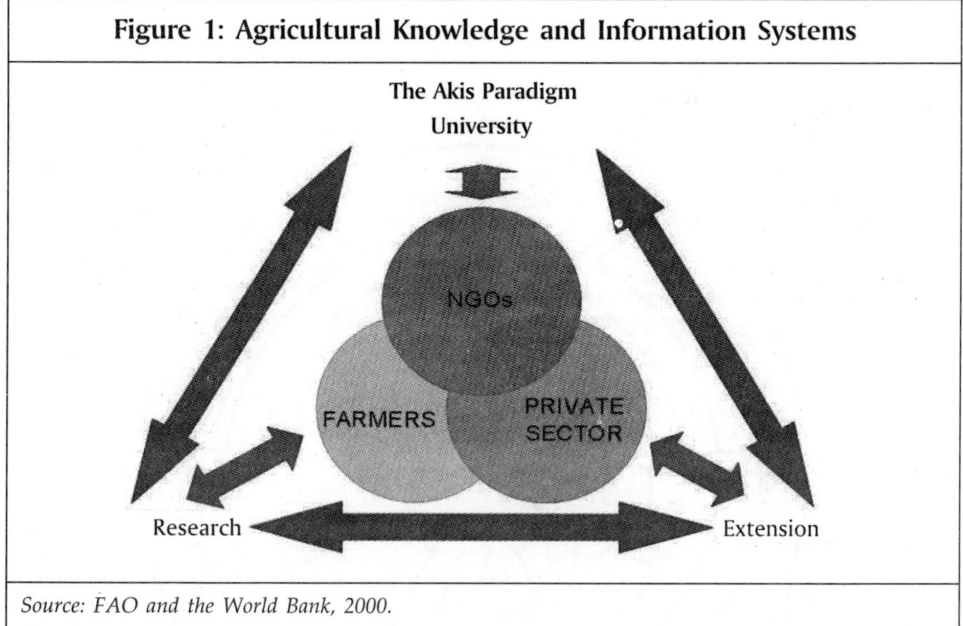

Source: FAO and the World Bank, 2000.

and Hall 2006).[2] Over the past few years, innovations have gained cachet in academic and donor circles just like Farming Systems Research (FSR) did in the late seventies and eighties (Collinson 1987). However, FSR research is now subsumed as part of participatory research. Likewise, FAO/Rome established a Sustainability Department in 1992 and then changed its name to a "Natural Resource Department" in January 1, 2007. Why? Several observers note that the sustainability paradigm never gained intellectual legitimacy and donor support. Will the innovation systems paradigm follow the same pathway as FSR and the FAO's experience with sustainability.

Innovation systems are now used as the centerpiece to develop a new generation of donor projects (Figure 2). Many countries are using enhanced innovation capacity as a "pipeline to donor funding." However, recent reports by (FARA 2007), and the World Bank point out the difficulties in operationalzing the innovation approach. It is an open question whether the innovation approach can be operationalized and be of much help in identifying how to promote the growth of smallholder agriculture in Africa. Although the AKIS paradigm is

[2] Nelson (1993), an economist, used the paradigm to explain the development of innovations that have spurred economic growth in industrial nations.

Figure 2: A Stylized Innovation System

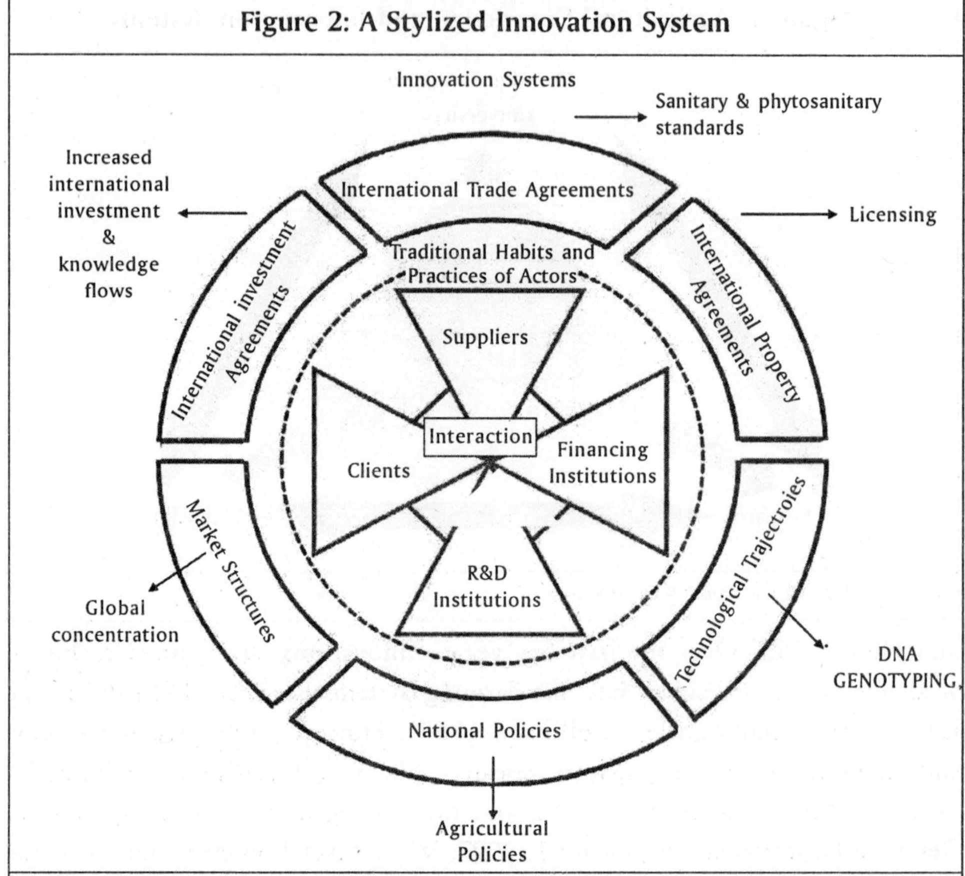

Adapted from: Lynn K. Mytelka, "Local Systems of Innovation in a Globalized World Economy" in Industry and Innovation, Vol. 7. No. 1, June 2000.

simplistic, it is still a useful concept in highlighting the importance of developing a *system* of education, technology and extension institutions that are the drivers of transforming traditional agriculture (Bonnen 1998). These three core institutions in the AKIS paradigm need to be interlinked and largely developed with public funding.[3] The challenge is for poor countries to invest in these core institutions to "buy their way into the growth process" up to a point where public-private partnerships and private investors, including farmers contribute the lion's share of the investment necessary to drive the agricultural transformation process (Evenson 2004).

[3] In India today roughly 97 percent of its extension budget comes from public sources.

Six Models of Agricultural Extension

Currently, there are six basic extension models in various stages of development and implementation in the developing world. Instead of trying to identify the "best fit" extension model for a particular country, the reality is that a pluralism of models is being used in most countries in Asia and Africa (Davis 2006; Birner and Anderson 2007 and Birner *et al.*, 2006). Virtually every developing country now has a mixture of public, NGO and private firms (e.g., seed and fertilizer dealers) delivering extension assistance to smallholders. The following six extension models are being used in developing countries.

1. *The national public extension model:* Thas been historically the dominant extension model throughout the world and it has usually been a key institution within and reporting to the Ministry of Agriculture. However, in the United States, public extension is located in the State Land Grant Universities. The transaction costs of the Land Grant model are low because one administrator, the Dean of Agriculture, is responsible for the coordination and management of three interlinked institutions: agricultural research, extension and agricultural higher education. Although India's State Agricultural University (SAU) Model is based on the US Land Grant Model, the SAUs are responsible to the Department of Agriculture in their respective states and the Indian Council of Agricultural Research.[4]

2. *The commodity extension and research model:* This was introduced by colonial powers in Malaysia, Mali and other colonies exporting cotton, palm oil etc. (Ruttan 1982; Eicher 1989). The model combines research and extension and it is still in operation in many countries today. In Mali, for example, smallholder cotton farmers are served by a self-financed cotton research and extension system while the public extension model serves farmers outside the cotton zone.

3. *The Training and Visit (T&V) extension model:* This was launched in Turkey in the early seventies and then spread to India and throughout Africa under World Bank sponsorship in the late seventies and the eighties. The T&V

---

[4] Since agriculture is a state subject in India, the SAUs in India are currently receiving approximately 80% of their funding from the state governments and 20% from the national government through the Indian Council for Agricultural Research.

model consumed about three billion dollars of donor assistance over the 1975-1995 period. However, the T&V model has proven to be financially unsustainable (Anderson, Feder and Ganguly 2006). Nevertheless, some countries (e.g., Zambia and Mali) are currently using modified T&V extension programs.

4. *The NGO extension model:* In the nineties, many NGOs shifted gears and moved from providers of food aid and humanitarian assistance to become "agents of development." The NGOs established food and community development projects in many African countries in the 1990s that were primarily financed by bi-lateral donors. For example, in Mozambique in 2005, the NGOs employed 840 extensionists as compared with 770 public extension workers (Gemo, Eicher and Teclemariam, 2005).

5. *The private extension model:* This is spreading in industrial countries such as the Netherlands, New Zealand, the United States and more recently in some middle income countries such as Chile and low income countries such as Uganda. Under the private model, the farmer is expected to pay some of the cost of extension with the hope that public outlays on extension will be reduced (Anderson and Crowder 2000). But there is little evidence to date that small scale farms can "buy their way out of poverty" by paying for extension advice. Several researchers are documenting the privatization of extension in Uganda but the jury is still out on the financial sustainability of private extension (Anderson 2007).

6. *The Farmer Field School (FFS) approach (model):*[5] This model emerged in Asia in the 1980s when extension workers offered advice to farmers on using IPM (Integrated Pest Management) to control pests in rice mono-cropping areas in the Philippines and Indonesia (Feder, Murgai and Quizon 2004a; Gallagher *et al.,* 2006). The model was remarkably effective in reducing pesticide use by up to 80 percent on farms in these two countries. The FFS model is now being used in around 50 developing countries. But farmers completing a school are reported to have limited success in spreading the new technology to their neighbors. Thus explains why there is a need for research on the following issues: do the field schools

[5] There is spirited debate among extension experts whether the FFS is an approach or a model.

increase the knowledge of farmers in the short-, medium- and long- run? Does the increased knowledge of farmers who have completed a school lead to higher crop yields and increased agricultural productivity? Why is the spread of technology from farmers attending school to neighboring farmers so limited? Finally, is the model financially sustainable?

We shall now turn to a discussion of the rise and fall of the T&V extension model and the growth of the Farmer Field School model that is now being implemented at sub-national and national levels in some 50 to 70 developing countries. We shall also highlight the emergence of the ATMA Extension model that is expanding rapidly in India and comment on China's new role in agricultural extension in Africa.

The T&V (Training and Visit) Extension Model

The Food and Agriculture Organization of the United Nations (FAO) provided global leadership in extension from the 1940s to the 1980s by drawing on its world-wide field experience and offering counsel to member states on a range of extension models and issues such as extension/farmer ratios and the performance of different extension models. But FAO's global leadership in extension was challenged in the late seventies and throughout the eighties by the World Bank's aggressive promotion and funding of a new extension model – the T&V model.

The T&V model was first developed in Turkey and then introduced in India in the 1970s and adopted by several dozen countries in Africa in the eighties. The premise of the model was that locally available technology was awaiting adoption by farmers. Therefore, subject matter extension specialists would visit with a group of "contact farmers" from surrounding villages every fortnight (later every month) and train them to take messages such as improved agronomic practices back to farmers in their villages. The T&V system prohibited front line extension agents from selling seed and fertilizer because of the belief that extension agents should focus and concentrate on introducing new food crop messages. But as more village extension agents were hired, costs mounted for state governments.

In practice, the T&V system was effective in disseminating Green Revolution technology, especially in the high-potential, irrigated areas in Asia, but it did not reach farmers in rain fed areas in Asia and Africa. Close to 50 developing countries utilized some form of T&V extension during the period from 1975-1995. However, the World Bank withdrew its support for the T&V model following a report that its high recurrent costs could not be reduced (Purcell and Anderson 1997). In India, for example, the introduction of T&V extension greatly expanded the number of Village Extension Workers (VEWs) in the State Departments of Agriculture, resulting in long term financial obligations for state governments. Since the T&V model turned out to cost about 25 percent more than the traditional public sector extension systems that it replaced in Ministries of Agriculture, the extension debate shifted in the late eighties and nineties to a new model- Farmer Field School (FFS) (Anderson, Feder and Ganguly, 2006). But some valuable lessons have been learned from the T&V experience. First and foremost is the need to analyze the cost of new models, fiscal implications of an expanded scale, the degree of dependence on external funding and the likelihood of domestic political support to pay the recurrent costs of scaling up new models over time (Anderson, Feder and Ganguly 2006). Nevertheless, some countries such as Mali are currently using a modified version of T&V called "Block Extension" (Dembele, 2007).

The Farmer Field School Extension Model

The Farmer Field School (FFS) model is a community-based learning system that was introduced in Asia in the eighties as an imaginative response to the overuse of insecticides in irrigated rice fields in Asia in the wake of the Green Revolution (Gallagher *et al.,* 2006). Farmers in the Philippines and Indonesia attended weekly meetings and taught themselves how to control insect damage. The FFS model is an example of group-based experiential learning (or "learning by-doing") that encourages farmers in "informal schools" to meet once a week in the same farmer's field and analyze and discuss their farming operations and then determine which agricultural interventions should be adopted and evaluated on their own farms. Normally, 20 to 30 neighboring farmers gather for group study on a member's farm once a week for about 14 weeks in a typical growing season.[6]

---

[6] Group learning among farmers is more than a century old in the Netherlands. Small flower producers held "winter evening meetings" exchanged experience and borrowed different types of greenhouses and green house glass from Denmark, the Chanel Islands and England. These human capital investments that laid the foundation for its present global flower industry (Eicher, 2006).

In East Africa, FFS networks, associations and federations have emerged that are farmer-owned and financed (Braun 2006). To date, Farmer Field Schools have turned out about 4 million graduates. The FFS model has facilitated the spread of Integrated Pest Management (IPM) practices in Asia over the past 15 years, and more recently in Africa. But there are some hard questions and more time are needed to answer the central question: is it a cost-effective learning innovation? We know that the model is more expensive than the traditional extension model where one extension worker can serve farmers via radio, and newspapers, etc. The critical questions are whether the model yields higher returns to pay for its added cost and can it be financed locally after foreign aid is phased out.

What is the empirical record of the FFS model? Four recent studies illustrate why FFS is an attractive model and why there is a need for more research on the short-, medium- and long-term impact of the model. The Sri Lanka Department of Agriculture, with support from FAO and a number of donors, ran an IPM program in Sri Lanka from 1995 to 2002 that included 610 FFS projects throughout the country. Tripp, Wijeratne and Piyadasa (2005) carried out a survey of FFS in southern Sri Lanka and found that FFS farmers growing rice who adopted FFS knowledge derived from IPM practices were able to reduce the number of applications of insecticides by 81 percent. But surprisingly, farmers completing the FFS did not adopt other recommended farm practices and the study provided little evidence of farmer to farmer transmission of the principal practices of the FFS. The authors of the Sri Lanka evaluation have called for more rigorous impact assessment because:

> *The insufficient assessment of FFS programs (and their alternatives) is a significant part of the problem. The FFS approach makes a very attractive package for donors and NGOs. It offers a well-defined subject introduced through a specific methodology. Courses and participants can be counted. Enthusiastic participants can be relied on to give glowing testimonials. As these experiences accumulate, an impression develops of FFS as a practical and widely applicable strategy, and while donors are unclear about objectives, and hence disorganized in their attempts at evaluation, FFS expands into new areas and makes new claims (Tripp, Wijerante and Piyadesa 2005).*

The Global IPM facility recently commissioned two experienced field researchers, Van den Berg and Jiggins (2007), to prepare a background paper on the state of the art of published and unpublished studies of the impact of FFSs on IPM in Asia. The authors stated their challenge as finding "a form of adult education that would capacitate the millions of smallholders to become experts in decentralized pest management through practical, field-based learning methods" (Van den Berg and Jiggins 2007, p.664) The authors admitted that the cost effectiveness of the Farmer Field Schools programs is a matter of "energetic debate" and that the results of many FFS studies reveal that the methodology for impact evaluation is "still under development." The findings of this valuable survey paper by van den Berg and Jiggings are the following:

- The evaluation of the FFS model combines pest management (IPM), new technology and farmer education makes it difficult to develop methodologies to study the impact of both of these activities over time.
- Most impact studies of FFS have concentrated on measuring immediate impacts, most notably the effects of insecticide use on crop yields. However, this type of methodology is weak for estimating medium- and long-term impacts such as developing social capital to build producer organizations.
- The opportunity cost (income forgone) of farmers attending weekly or bi-weekly or monthly FFS meetings should be taken into consideration as a cost issue. In a study of cotton in Mali, it was found that the *annual* opportunity cost of a farmer's time in attending 14-20 weekly sessions on cotton was US$20. This is a cost that should be included in cost-benefit calculations.
- The immediate impact of FFS on farmers producing rice in Asian countries is the reduction in pesticide use while the achievement of FFS on other continents "remains to be established."
- FFS programs in Asian countries have only covered one to five percent of all farm households (Van den Berg and Jiggins (2007).

Sierra Leone recently launched an ambitious food security program called "Operation Feed the Nation." After a decade of Civil War, the President of Sierra Leone pledged his support for this program so that "within five years, no Sierra

Leonean should go to bed hungry." The FAO was invited to help oversee a quick study of the 510 Farmer Field Schools. The study was carried out by the Overseas Development Institute (ODI) and Dunstan Spencer and Associates in early 2006. The study was carried out in three districts over two months and it found that:

- The results of the evaluation were positive but the authors concluded that the overall impact of the FFS cannot be known for certain because of the lack of reasonably accurate baseline data for comparison.
- Impact monitoring is currently lacking.
- The FFS costs between US$ 16 and US$ 47 per farmer served for each cycle or season. (Longley, Spencer and Wiggins, 2006).

In Kenya, a recent FAO commissioned study reports that Farmer Field School (FFS) Networks emerged in Western Kenya during 2000 as a result of exchange visits and communication between farmers, facilitators, trainers and project staff (Braun 2006). Similar networks have subsequently emerged elsewhere in Kenya, Uganda and Tanzania. These FFS Networks were formed by farmers who graduated from an FFS. FFS networks in Western Kenya have shown how farmers themselves have been able to build bottom-up producer organizations during and after the completion of donor projects. This self-emergence of FFS networks depicts FFS as an effective approach to organize and empower farmers.

To summarize, the Farmer Field School model is an important institutional innovation that needs to be studied in depth in different agro-ecological zones, different institutional arrangements and over time. Because of the lack of baseline data and adequate monitoring of ongoing FFS activities at the farmer and community levels, the available evidence suggests that it is premature to promote the FFS model as the "best model" for developing countries.[7] It will take time and resources for researchers to study and evaluate this important institutional innovation. Meanwhile, international organizations such as the FAO should endorse the concept of the pluralism of extension models, including the FFS models. Countries should be encouraged to collect data on the impact, costs and returns of the FFS model, including its financial sustainability, in a learning-by-doing manner.

---

[7] Davis (2006) surveyed the spread of FFRs in Africa and concluded that FFSs do not represent a "silver bullet" for Africa. However, Gallagher *et al.*, 2006 challenged the survey because of an alleged "inadequate sampling of evidence."

Another area that warrants ongoing attention by researchers is the optimal size of national extension systems, including government, private and NGOs. Simply increasing the number of extension agents as the engine of agricultural development was found to be unsuccessful in Latin America in the forties and fifties (Rice, 1971). Later, African governments added 57,000 extension agents to their government extension departments from 1959 to 1980 (Judd, 1987). Today, India has around 100,000 extension agents and China has around 140,000. Ethiopia has around 10,000 extension agents and it is training an additional 30,000 agents even though there are many reservations about the fiscal sustainability of such a large build up. Clearly there is a need for an expanded research program on alternative extension models in developing countries. Yet research on extension is chronically under-funded (Anderson, 2007).

New Extension Models

The Agricultural Technology Management Agency Model (ATMA)

One of the recurring criticisms of national public extension systems is that they are highly centralized and they inhibit the feedback from clients to extension specialists, researchers, policy makers and donors. Decentralization of extension to local governments proceeded rapidly in many Latin American countries in the 1980s and 1990s. Decentralization is now underway in Asia.[8] In India, decentralization is being pursued through a new model of extension – the Agricultural Technology Management Agency Model (ATMA), an autonomous organization that was initially set up in the late 1990s with World Bank support (Singh, Swanson and Singh 2006). The ATMA model (Figure 3) combines decentralization with a focus on agricultural diversification and increasing farm incomes and employment. Decisions on extension are made by a Governing Board with equal representation between:

- The heads of the line departments, including animal husbandry, horticulture, etc. and key people in the State Department of Agriculture
- Research units within the districts and stakeholder representatives ar
- A cross-section of farmers, women, disadvantaged groups and the private sector.

[8] Van den Ban and Samantha, (2006).

Figure 3: Agricultural Technology Management Agency (ATMA)

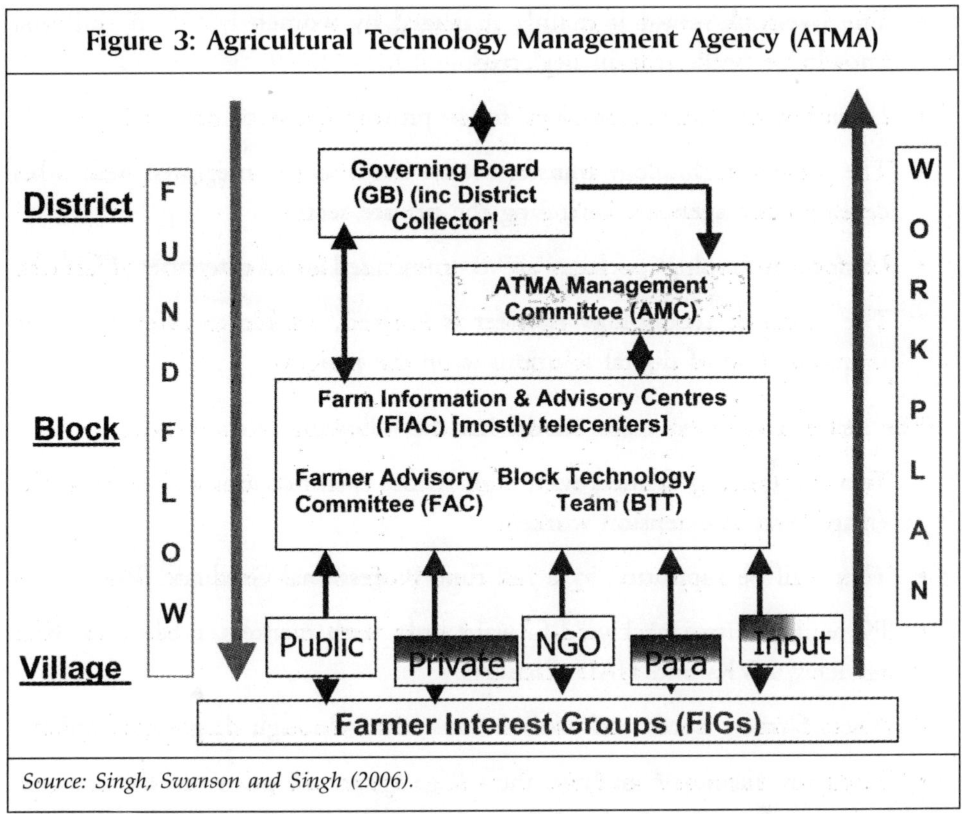

Source: Singh, Swanson and Singh (2006).

The ATMA model is considered a major success in India because after only five years of operation (2001 to 2006), it has been adopted in 60 districts (about 10% of all districts in India). The model is slated to be extended to all 600 districts within the next five years (Anderson 2007 citing Swanson 2006, p. 14).

India: The New Haryana Extension Model

The state of Haryana and the neighboring state of Punjab are considered the breadbaskets of India. But the Vice Chancellor of Haryana Agricultural University – J C Katyal (2007) reports that the present extension model is in need of reform. Therefore, Haryana University is experimenting with a new extension model that deals with what the Vice Chancellor describes as the "stress points" of the present extension system:

- The technology transfer service does not cover physical, social and economic aspects of an integrated farming system.

- The livestock sector is mainly managed by women but their skill and knowledge needs remain neglected.
- Extension has little technology for improving livestock productivity.
- The present technology transfer model needs to be integrated with other development agencies, including the private sector.
- Uniform technology packages are recommended for all categories of farmers.

  The reach of technology transfer is limited, unable to offer real time solutions. Use of digital solutions is on the margins.

The new Haryana extension model has the following characteristics:

- Two enterprising farmers with high school qualifications will work at the village level as extension workers.
- They will be supported by a full time Professional-Graduate (PG).
- PG will facilitate backward-forward links with markets, credit and other agencies, KVKs and HAU scientists.
- Assess farmers needs and analyze constraints through diagnostic studies.
- Based on diagnostic analysis, the village agent will prioritize the activities in association with the local farmers.

The new Haryana model will be tested and applied in two representative villages selected from each of the 20 districts of Haryana.

China's Approach to Agricultural Extension in Africa

President Hu Jintao of China identified agricultural cooperation as one of eight types of technical assistance to Africa at the China-Africa summit in November 2006. The Chinese Ministry of Commerce and Ministry of Agriculture has jointly sent five working groups to 14 African countries to investigate setting up ten agricultural technology demonstration centers in Africa and supplying 100 senior agro-tech experts to assist with Africa's agricultural development.

A training course for African officials on the extension of agricultural technology was held by the Chinese Ministry of Agriculture in Beijing in July 2007. A total of 35 agricultural officials from 21 African countries attended the training course, which was billed as the "2007 International Agro-Tech Extension Seminar for Africa." The course included lectures on genetically modified cotton and seed production technologies and the use of water-saving and biological technologies in agriculture.

Smallholders' Main Source of New Technology

The critics of public extension often charge that it reaches only a small percentage of smallholder farmers. To address this issue, we turn to the results of a 2003 national survey of 51,770 farm households in India who were asked to reveal their main source of information about new technology and farm practices over the past 365 days.

Figure 4 shows that the progressive farmers were the most important source (16.7%) of information for smallholders over a period of 12 months, followed by input dealers, radio and television. Surprisingly, only 6% of the farmers in the national survey gained their information from extension workers.

Figure 4: India – Percentage of Farm Households Accessing Information on Modern Agricultural Technology Through Different Sources, 2003

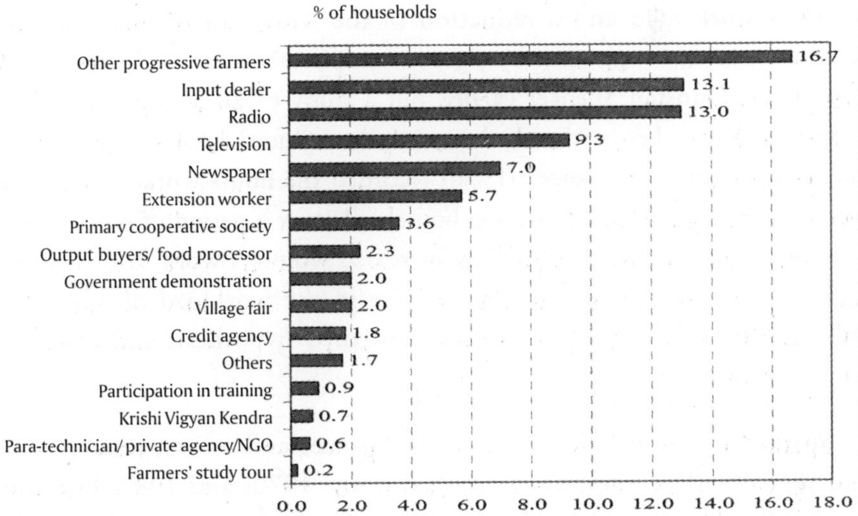

% of households

Source	% of households
Other progressive farmers	16.7
Input dealer	13.1
Radio	13.0
Television	9.3
Newspaper	7.0
Extension worker	5.7
Primary cooperative society	3.6
Output buyers/ food processor	2.3
Government demonstration	2.0
Village fair	2.0
Credit agency	1.8
Others	1.7
Participation in training	0.9
Krishi Vigyan Kendra	0.7
Para-technician/ private agency/NGO	0.6
Farmers' study tour	0.2

0.0 2.0 4.0 6.0 8.0 10.0 12.0 14.0 16.0 18.0

Source: Derived from data reported in NSSO (2005: 7).

To be sure, this finding can be used to call for increased privatization of extension. But, closer examination reveals that extension is reaching a small percentage of smallholder farmers because extension is dramatically under funded. For example, India's Extension Policy Framework reports that "States have barely been able to pay the salaries of extension personnel. Less than 10% of the budget is available for operational expenses, which has practically immobilized the service with scarcely any technology dissemination in the field" (Birner and Anderson 2007). The lesson of this national survey of 51,770 farmers reveals that India's public extension system is not adequately financed. Rather than calling for privatization, of extension, new public-financed extension models (such as ATMA) should be piloted over the coming five to seven years.

Needed Research on Woman's Role in Agricultural Development and Extension

The literature on the role in extension in helping women make decisions on the adoption of improved farm practices is rather thin even though women, especially in Africa, are major producers of food crops and active and shrewd traders in local markets. When Ester Boserup published her path-breaking book *Woman's role in Economic Development* (1970), she charged that women "lose in the development process" because agricultural development projects can lead to an increase in women's work load and a reduction in the workload of men. But Boserup's assertion was not supported by rigorous empirical research. To test the Boserup hypothesis, Dunstan Spencer carried out a study of an agricultural development project in Sierra Leone that had recently introduced land-saving biological and chemical technology. Spencer (1976) recorded the hours worked per day by men, women and children in 23 farming households over a period of one year (1974/75) and found that the new technology increased women's work load slightly but the increase was much less than the increase in the workload of adult males and children. Therefore, Spencer rejected Boserup's hypothesis and called for more empirical research.

Sparked by the findings of Boserup, Spencer and others, research on women and agricultural development flourished in the 1970s and 1980s but dried up in the 1990s. The World Bank's poverty reduction mandate generated a 1994 Operational Policy on Gender Dimensions that has produced a string of recent books on gender issues (Gopal 2005; Ellis *et al.*, 2006 and Ellis, Manuel and Blackden 2007). However, these books gloss over women's role as farmers, traders and participants in rural non farm activities. The bottom line is that little progress has been made in figuring out how to assist rural women increase their voice, and their profit in farming and related industries.

Measurement Issues

One of the most complex issues for governments and donors is developing a capacity to measure the performance of alternative extension models. Feder, Willett and Zijp (2001) have identified eight generic issues that explain why there is so much confusion about appropriate measurement techniques and the performance of various extension models that developing countries should pursue. These generic factors are as follows:

1. *Difficulty in attributing impact:* Attribution of impact of extension programs is an analytically challenging task due to the lack of baseline information, unavailability of appropriate control groups, and the systemic biases in extension placement and contracts (Birkhaeuser, Evenson and Feder, 1991).
2. *Scale and Intensity:* The cost of reaching large, geographically dispersed and remote smallholder farmers is high, particularly given high levels of illiteracy, limited access to mass media, and high transport costs.
3. *Dependence on broader policy environment:* The outcome of extension efforts depends on policies over which agents and their managers have little influence (input and output prices, credit policies, input supplies, marketing and infrastructure system).
4. *Interaction with knowledge generation:* Public extension and research systems often compete for budgets, but research institutions often have an advantage because of their higher status, better management quality, and links with the global science community.
5. *Weak accountability:* Weak accountability (linked to the inability to attribute impact) is reflected in low-quality and repetitive advice given to farmers, and in diminished effort to interact with farmers, and to learn from their experience.

6. *Weak political commitment and support:* Many donors report that highly visible irrigation or road projects are often more attractive to politicians than extension expenditures.

7. *Public duties other than knowledge transfer:* Governments often use field-level cadres of public servants who are already present in rural areas for nonextension duties such as collection of statistics, distribution of subsidized inputs, assisting and collecting loan applications, and election campaign work on behalf of local or national ruling parties.

8. *Financial un-sustainability:* This can cause the demise of any investment program, but it is observed that at times of general fiscal constraints, extension budgets are more likely to be squeezed due to weak political support (Feder, Willet and Zijp, 2001).

Summary

One of the highlights of this review is the vast output of the, papers and articles on extension reforms in Asia relative to Africa. For example, the recent book *Changing Roles of Agricultural Extension in Asian Nations* (Van Den Ban and Samanda 2006) is a comprehensive and valuable survey of extension reforms underway in Asia. There is a need for a similar book on extension reforms in Africa. Socio-economic research is urgently needed on the following topics.

- Comparative studies of the FFS model in Asia and Africa.
- Fast track programs to increase farmer extension coverage.
- A study of the ATMA model in India and its lessons for Africa.
- The role on ICTs in reducing the cost of supplying information to farmers.
- Development of impact assessment techniques to measure the short, medium and long term impacts and financial sustainability of Farmer Field School extension programs.
- Case studies of new models of public and private cooperation in delivering extension to farmers.
- Case studies of developing national market information systems in Asia and Africa.

- What can Africa learn from the global experience in increasing the number of women employed as extension managers and front line extension agents?
- Case studies of scaling up successful NGO extension projects.
- Role of private companies delivering inputs and providing extension assistance to farmers in Asia: Lessons for Africa.

(Carl K Eicher, University Distinguished Professor, Department of Agricultural Economics, MSU, East Lansing, Mi: The author can be reached at ceicher@msu.edu).

References

Anandajayasekeram, P. ed. (2007). *Impact of Science on African Agriculture and Food Security.* Cambridge, Ma.: CABI Books.

Anandajayasekeram, P. K Davis, S. Workneh. (2007). "Farmer Field Schools: An Alternative to Existing Extension Systems? Experience from Eastern and Southern Africa". *Journal of International Agricultural and Extension Education* 14(1): 81-93.

Anderson, Jon and L. Van Crowder. (2000). "The Present and Future of Public Sector Extension in Africa: Contracting-out or Contracting-in?" *Public Administration and Development.*

Anderson, Jock. (2007). "Agricultural Advisory Services. A background paper for WDR 2008". Washington, D.C.: World Bank.

* Anderson, Jock R., and Gershon Feder. (2007). "Agricultural Extension." *In* Robert E. Evenson and Prabhu Pingali (eds.) *Handbook of Agricultural Economics.* Amsterdam: Elsevier.

* Anderson, Jock R., Gershon Feder and Sushima Ganguly. (2006). "The Rise and Fall of Training and Visit extension: An Asian mini-drama with an African Epilogue?" In A.W. Van den Ban and R.K. Samanta eds., *Changing Roles of Agricultural Extension in Asian Nations,* New Delhi: B.R. Publishing Company: 149-172.

Antholt, Charles H. (1998). "Agricultural Extension in the Twenty-first Century". In *International Agricultural Development.* Third edition, eds. Carl K. Eicher and John M. Staatz. Baltimore, MD.: The John Hopkins University Press: 354-369.

Arunachalam, S. (2002). "Reaching the Unreached: How Can We Use ICTs to Empower the Rural Poor in the Developing World through Enhanced Access to Information?" Paper presented at the 68th International Federation of Library Associations (IFLA) Council and General Conference. Glasgow.

* Publications of enduring value.

Asuming-Brempong, S., D. B. Sarpong, and F. Asante. (2006). "Institutional Bottlenecks of Agricultural Sector Development: The Case of Research and Extension Provision in Ghana." University of Ghana for the OECD Development Centre (Paris, France). Accra: Ghana. Processed.

Atchorarena, D. and L. Gasperini. (2003). *Education for Rural Development: Towards New Policy Responses.* Rome and Paris: FAO and UNESCO.

Benor, D., J. Harrison and M. Baxter. (1984). *Agricultural Extension: The Training and Visit System.* World Bank, Washington, D.C.

Bindlish, V. and R. Evenson. (1997). "The Impact of T&V Extension in Africa: The Experience of Kenya and Burkina Faso". *World Bank Research Observer* 12(2): 183-201.

Binswanger, Hans P. (2007). "Empowering Rural People for their own Development". In *Contributions of Agricultural Economics to Critical Policy Issues.* Keijiro Otsuka and Kaliappa Kalirajan, editors. Malden, MA: Blackwell.

Birkhaeuser, D., R. E. Everson and G. Feder. (1991)."The economic impact of agricultural extension: A review", *Economic Development and Cultural Change* 39, 607-50.

Birner, Regina and Jock R. Anderson. (2007). "How to Make Agricultural Extension Demand -Driven? The Case of India's Agricultural Extension Policy". Washington, D.C.: ISNAR/ INPRI, Draft.

Birner, Regina, Kristin Davis, John Pender, Ephraim Nkonya, Ponniah Anandajayasekeram, Javier Ekboir, Adiel Mbabu, David Spielman, Daniela Horna, Samuel Benin, and Marc J. Cohen. (2006). "From 'Best Practice' to 'Best Fit': A Framework for Analyzing Pluralistic Agricultural Advisory Services Worldwide." Washington, D.C.: International Food Policy Research Institute, DSGD Discussion Papers 37. Available on line at *http://www.ifpri.org/DIVS/DSGD/dp/dsgdp37.asp*

Bonnen, James T. (1998). "Agricultural Development: Transforming Human Capital, Technology, and Institutions". In *International Agricultural Development,* Third Edition, C.K. Eicher and J.M. Staatz eds. Baltimore: Johns Hopkins University Press: 271-286.

Boserup, Ester. (1970). *Woman's Role in Economic Development.* New York: St. Martins's Press.

Box, Louk and Rutger Engelhard. Eds. (2006). *Science and Technology Policy for Development: Dialogues at the Interface.* London: Anthem Press.

Braun, Arnoud. (2006). "Farmer Field School Networks in Western Kenya: Evolution of Activities and Farmers' Priorities". Rome: FAO. Draft.

Byerlee, Derek, Xinshen Diao and Chris Jackson. (2006). *Agriculture, Rural Development, and Pro-poor Growth: Country Experiences in the Post-Reform Era.* Agriculture and Rural Development Discussion Paper 21. Washington, D.C.: World Bank.

Chipeta, Seena. (2006). *Demand Driven Agricultural Advisory Services.* Lindau: Neuchatel Group. Available on line at *http://www.neuchatelinitiative.net/english/documents/DemandDrivenAgriculturalAdvisoryServices.pdf*

Collier, Paul. (2007). *The Bottom Billion: Why the Poorest Countries Are Failing and What Can Be Done About It.* New York: Oxford University Press.

Collinson, M.P. (1987). "Farming Systems Research: Procedures for Technology Development". *Experimental Agriculture* 33:365-386.

Crane, Diane. (1972). *Invisible Colleges: Diffusion of Knowledge in Scientific Communities.* Chicago: University of Chicago Press.

* Crowder, L. Van and J. Anderson. (1997). "Linking Research, Extension and Education: Why is the Problem so Persistent and Pervasive?" *European Journal of Agricultural Education and Extension.* 4:241-249.

Dalsgaard, Jens, Peter Tang, Thai Thi Minh, Vo Ngan Giang, and Jens Christian Riise. (2005). "Introducing A Farmers' Livestock School Training Approach into the National Extension System in Vietnam". AgREN. Network Paper No. 144. January. London: ODI.

Davidson, A.P. and M. Ahmad. (2003). *Privatization and the Crisis of Agricultural Extension: The Case of Pakistan*, Aldershot: Ashgate.

Davis, Kristin.(2006). "Farmer field schools: A boon or a bust for extension in Africa". *Journal of International Agricultural and Extension Education.* 13(1):91-97.

Del Castello, Riccardo and Paul Mathias Braun. Eds. (2006). "Framework on Effective Rural Communication For Development". Rome: FAO and GTZ.

Dembele, Nango. (2007). "E-mail". May 26.

Eicher, Carl K. (1982). "Reflections on the Design and Implementation of the Senegal Agricultural Research Project". East Lansing, Department of Agricultural Economics, Michigan State University.

Eicher, Carl K. (1988). "An Economic Perspective on the Sasakawa-Global 2000 Initiative to Increase Food Production in Sub-Saharan Africa". Paper presented at the workshop "Reviewing the African Agricultural Projects." The Sasakawa-Global 2000 Africa Initiative, Nairobi, 18 March. Atlanta, Georgia: The Carter Center.

* Eicher, Carl K. (1989). *Sustainable Institutions for African Agricultural Development.* Working Paper No. 19. The Hague: ISNAR.

Eicher, Carl K. (2004). "Rebuilding Africa's Scientific Capacity in Food and Agriculture". Background Paper No. 4 commissioned by the InterAcademy Council (IAC) Study Panel on Science and Technology Strategies for Improving Agricultural Productivity and Food Security in Africa. Amsterdam: InterAcademy Council. Available at *www.interacademycouncil.net*

Eicher, Carl K. (2006). "The Evolution of Agricultural Education and Training: Global Insights of Relevance for Africa". Staff Paper 2006-26. East Lansing, MI, Department of Agricultural Economics, Michigan State University. Available on line at *http://agecon.lib.umn.edu/cgi-bin/pdf_view.pl?paperid=23691&ftype=.pdf*

Eicher Carl K. and Lawrence Witt. eds. (1964). *Agriculture in Economic Development.* New York: McGraw-Hill.

* Eicher, Carl K. and John M. Staatz. Eds. (1998). *International Agricultural Development.* Third edition. Baltimore. Md.: The Johns Hopkins University Press.

Eicher, Carl K. and Mandivamba Rukuni. (2003). *The CGIAR in Africa: Past, Present and Future.* (Thematic Working Paper). Washington D.C.: World Bank Operations Evaluation Department. Available at *www.worldbank.org/oed/cgiar.*

Eicher, Carl K., K. Maredia, and I. Sithole-Niang. (2006). "Crop biotechnology and the African farmer". *Food Policy* 31 (6): 504-527.

Ellis, Amanda, Claire Manual and Mark C. Blackden. (2005). *Gender and Economic Growth in Uganda: Unleashing the Power of Women.* Washington, D.C.: World Bank.

Ellis, Amanda *et al.*,. (2007). *Gender and Economic Growth in Kenya: Unleashing the Power of Women.* Washington, D.C.: World Bank.

Evenson, Robert E. (2004). *Making Science and Technology Work for the Poor: Green and Gene Revolutions in Africa.* (UNU-INRA Annual Lectures, 2003) Accra and Addis Ababa: UNU-INRA and UN ECA.

FAO. (1974). *Agricultural Extension: A Reference Manual.* Rome: FAO.

FAO. (2006). "Building A Virtual Extension and Research Communication Network (VERCON): An Action Oriented Guide. Rome: FAO".

FAO and the World Bank. (2000). *Agricultural Knowledge and Information Systems for Rural Development (AKIS/RD): Strategic Vision and Guiding Principles.* Rome and Washington: FAO and World Bank.

FARA. (2007). *FARA's 2007-2016 Strategy: Enhancing African Agricultural Innovation Capacity.* Accra: FARA.

Farrington, J. (1994). "Public sector agricultural extension: Is there life after structural adjustment?" *Natural Resource Perspectives* 2. ODI, London.

Feder, G., R. E. Just and D. Zilberman. (1985). "Adoption of Agricultural Innovations in Developing Countries: A Survey". *Economic Development and Cultural Change* 33(2): 255-98.

Feder, Gershon and Roger H. Slade. (1993). "Institutional Reform in India: The Case of Agricultural Extension". In K. Hoff, A. Braverman, and J.E. Stiglitz, eds., *The Economics of Rural Organization: Theory, Practice and Policy.* New York: Oxford University Press: 530-542.

Feder, Gershon and Sara Savastano. (2006). "The Role of Opinion Leaders in the Diffusion of New Knowledge: The Case of Integrated Pest Management". *World Development* 34(7): 1287-1300.

Feder, Gershon, Rinku Murgai and Jaime B. Quizon. (2004a). "Sending Farmers Back to School: The Impact of Farmer Field Schools in Indonesia". *Review of Agricultural Economics.* 26(1): 45-62.

Feder, Gershon, Rinku Murgai and Jaime B. Quizon. (2004b). "The Acquisition and Diffusion of Knowledge: The Case of Pest Management Training in Farmer Field Schools, Indonesia". In *Journal of Agricultural Economics.* 55(2):217-239.

Feder, Gershon, A. Willett, and W. Zijp. (2001). "Agricultural Extension: Generic Challenges and the Ingredients for Solutions". In S. Wolf and D. Zilberman, eds., *Knowledge Generation and Technical Change: Institutional Innovation in Agriculture.* Boston, Mass, Kluwer.

Gallagher, Kevin, Arnoud R. Braun and Deborah Duveskog. (2006). "Demystifying Farmer Field School Concepts". *Journal of International Agricultural and Extension Education.*

Garforth, Chris. (2005). "The Challenges of Agricultural Extension". In *Starter Packs: A Strategy to Fight Hunger in Developing Countries?* Edited by S. Levy. Cambridge, MA: CABI Publications.

Gemo, Helder, Carl K. Eicher and Soloman Teclemariam. (2005). *Mozambique's Experience in Building a National Extension System*, East Lansing, Mi.: Michigan State University Press.

Gopal, Gita. (2005). *Evaluating a Decade of World Bank Gender Policy: 1990-1999.* Washington, D.C.: World Bank.

Govereh, J., J. Shawa, E. Malawo, and T. S. Jayne. (2007). "Raising the Productivity of Public Investments in Zambia's Agricultural Sector", Working Paper 20, Food Security Research Project, Lusaka, Zambia.

Goyal, Apartija. (2006). "The Role of Information and Direct Access to Buyers: Evidence from Rural Agricultural Markets in Central India." University of Maryland. College Park, M.D. Processed.

Haggblade, Steven. (2007). "Returns to Investment in Agriculture. Food Security Research Project-Zambia". Ministry of Agriculture & Cooperatives Policy in Thesis number 19. Lusaka, Zambia: Michigan State University and GART.

Holdcroft, Lane. (1984). "The Rise and Fall of Community Development, 1950-65: A Critical Assessment". In *Agricultural Development in The Third World.* Carl K. Eicher and John M. Staatz, eds. Baltimore: Johns Hopkins University Press, pp. 46-58.

Hopper, W. David. (1968). "Investment in Agriculture: The Essentials for Payoff". In *Strategy for the Conquest of Hunger: Proceedings of a Symposium Convened by The Rockefeller Foundation, April 1-2, 1968 at the Rockfeller University.* New York: The Rockefeller Foundation.

Howell, J. (1983). "Strategy and Practice in the T&V System of Agricultural Extension", *Agricultural Administration Network Discussion* Paper No. 10, Overseas Development Institute, London.

Husain, S. S., D. Byerlee, and P. W. Heisey. (1994). "Impacts of the Training and Visit Extension System on Farmers' knowledge and Adoption of Technology: Evidence from Pakistan", *Agricultural Economics* 10(1): 39-47.

Judd, M. Ann, James Boyce and Robert Evenson. (1987). "Investment in Agricultural Research and Extension". In *Policy for Agricultural Research.* Edited by Vernon Ruttan and Carl Pray. Boulder, CO: Westview. pp. 7-38.

Katyal, J.C. (2007). "State of Technology Transfer: Proposal of a New Extension Model". CCS Haryana Agricultural University, Hisar, India.

Kremer, M. and A. P. Zwane. (2005). "Encouraging Private Sector Research for Tropical Agriculture". *World Development.* 33(1):87-105.

Labarta, Ricardo A. and Scott M. Swinton. (2005). "Do Pesticide Hazards to Human Health and Beneficial Insects Cause or Result from IPM Adoption? Mixed Messages from Farmer Field Schools in Nicaragua". Paper presented at the American Agricultural Economics Association Annual Meeting, Providence, Rhode Island, 24-27 July.

Lele, Uma. ed. (1991). *Aid to African Agriculture: Lessons from two Decades of Donors' Experience.* Baltimore: Johns Hopkins University Press.

Lele, Uma and Arthur A. Goldsmith. (1989). Development of National Agricultural Research Capability: India's Experience with the Rockefeller Foundation and its Significance for Africa. *Economic Development and Cultural Change* 37: 305-343.

Lele, Uma, *et al.*,. (2007). "Scaling Up Development Assistance: Lessons from Donor Evaluations and Evaluation Journals for Achieving Large Scale Sustainable Impacts". Washington, D.C. Draft. January 6.

Lio, Monchi, and Meng-Chun Liu. (2006). "ICT and Agricultural Productivity: Evidence from Cross-country Data." *Agricultural Economics* 34(3):221–28.

Longley, Catherine, Dustan Spencer and Steve Wiggins. (2006). *Assessment of Farmer Field Schools in Sierra Leone, Operation Feed The Nation.* London/Freetown: ODI and DS and A.

Moore, Mick. (1984). "Institutional Development, the World Bank, and India's New Agricultural Extension Programme". In *Journal of Development Studies.*

Mooris, Michael, Valerie A. Kelly, Ron J. Kopicki, and Derek Byerlee. (2007). *Fertilizer Use in African Agriculture: Lessons Learned and Good Practice Guidelines.* Washington, D.C.: The World Bank.

Muto, M. (2006). "Impacts of Mobile Phone Coverage Expansion and Roads on Crop Marketing of Rural Farmers in Uganda." Paper presented at the World Development Report Tokyo Consultation, ppt presentation. Tokyo, JBIC Institute.

Nagel, Uwe Jens. (1980). "Institutionalization of Knowledge Flows". *Quarterly Journal of International Agriculture.* Vol. 20, special issue.

Nahdy, Silim, Francis Byekwaso, and David Nielson. (2002). "Decentralized Farmer-Owned Extension in Uganda". In *Food Security in a Changing Africa.* Steven Breth, Editor, Geneva: Centre for Applied Studies in International Negotiations, pp 38-50.

Neuchatel Group. (2006). "Demand-Driven Agricultural Advisory Services". Lindau. *http://www.neuchatelinitiative.net/english/documents/DemandDrivenAgriculturalAdvisoryService s.pdf*

Nelson, Richard.(1993). *National Innovation Systems: A Comparative Analysis.* Oxford: Oxford University Press.

Nielson, David and Peter Bazeley. (2000). "The Search for Impact and whether 'Outreach' is the Answer." Nairobi: Kenya Agricultural Research Institute.

NSSO. (2005). "Situation Assessment Survey of Farmers – Access to Modern Technology for Farming". National Sample Survey, 59th Round (January – December 2003), Report No. 499(59/33/2), National Sample Survey Organization (NSS), Ministry of Statistics and Programme Implementation, Government of India, New Delhi.

* Pardey, Philip G. *et al.,* (2006). *Agricultural Research: A Growing Global Divide.* Washington, D.C.: IFPRI.

Perkinson, Ron. (2005). "Beyond Secondary Education: The Promise of ICT for Higher Education and Lifelong Learning". In *E-development: From Excitement of Effectiveness.* Robert Schware, ed. Washington, D.C.: World Bank.

Pingali, Prabhu. (2007). "Will the Gene Revolution Reach the Poor? – Lessons from the Green Revolution". Mansholt Lecture, Wageningen University, January 26.

Prahalad, C.K. (2006). *The Fortune At the Bottom of the Pyramid: Eradicating Poverty Through Profits.* Wharton School Publishing. Upper Saddle River: New Jersey.

Purcell, Dennis L. and Jock R. Anderson. (1997). *Agricultural Extension and Research: Achievements and Problems in National Systems.* Washington, D.C.: The World Bank Operations Evaluation Department.

Qamar, M. Kalim. (2006). "Privatization of Government Agricultural Extension Agencies". In *Changing Roles of Agricultural Extension in Asian Nations.* A.W. Van den Ban and R. K. Samanta eds. New Delhi: B.R. Publishing Corporation: 253-270.

Quarry, Wendy and Ricardo Ramirez. (2006). "A compendium of Regional Perspectives in Communication for Development". World Congress on Communication for Development. Rome: FAO. October 25-27.

Quizon, Jaime, Gershon Feder and Rinku Murgai. (2001). "Fiscal Sustainability of Agricultural Extension: The Case of The Farmer Field School Approach". *Journal of International Agricultural and Extension Education.* 8(1): 13-23.

Ramachander, Sangamitra and Asok Jhunjhunwala. (2006). "ICT and Agricultural Diversity". In *Changing Roles of Agricultural Extension in Asian Nations.* A.W. Van den Ban and R. K. Samanta eds. New Delhi: B.R. Publishing Corporation: 388-408.

Reardon, T. and C. P. Timmer. (2007). "Transformation of Markets for Agricultural Output in Developing Countries Since 1950: How Has Thinking Changed?" In R.E. Evenson, P. Pingali, and T.P. Schultz (eds) *Volume 3 Handbook of Agricultural Economics: Agricultural Development: Farmers, Farm Production and Farm Markets.* Amsterdam: Elsevierr Press: 2808-2855.

Rice, E. B. (1971). *Extension in the Andes.* Washington, D.C.: USAID.

Rivera, W. M. (1996). "Agricultural Extension in Transition Worldwide: Structural, Financial and Managerial Reform Strategies." *Public Administration and Development* 16:151-61.

Rivera, William M and John W. Carey. (1997). "Privatizing Agricultural Extension". In *Improving Agricultural Extension: A Reference Manual.* Edited by Burton E. Swanson, Robert P. Bentz and Andrew J. Sofranko. Rome: FAO, 203-211.

* Rivera, William M. and William Zijp. eds. (2002). *Contracting for Agricultural Extension: International Case Studies and Emerging Practices.* New York: CABI Publishing.

* Rogers, E.M. (2003). *Diffusion of Innovations,* (5th edition), New York. The Free Press, Macmillan Publishing.

Rola, A.C. J.B. Quizon and S.B. Jamias. (2002). "Do Farmer Field School Graduates Retain and Share What They Learn?: An Investigation in Iloilo, Philippines". *Journal of International Agricultural and Extension Education,* 5(1): 65-75.

* Roling, Niels. (1988). *Extension Science. Information Systems in Agricultural Development.* Cambridge: Cambridge University Press.

Roling, Niels. (2006). "Communication for Development in Research, Extension and Education", Thematic Paper no. 2, Communication Roundtable, 6-9 September, Rome: FAO.

Ruttan, Vernon. 1982. *Agricultural Research Policy.* Minneapolis: University of Minnesota Press.

Ruttan, Vernon. (2003). *Social Science Knowledge and Economic Development An Institutional Design Perspective*. Ann Arbor. Mi.: University of Michigan Press.

Saint, William S. (2003). Tertiary Distance Education and Technology in Sub-Saharan Africa. In D. Teferra and P. Altbach (eds.) *African Higher Education: An International Reference Handbook*. Bloomington, Indiana: Indiana University Press: 93-110.

Schultz, Theodore W. (1964). *Transforming Traditional Agriculture*. New Haven: Yale University Press.

Singh, J.P., Burton E. Swanson, K. M. Singh. (2006). Developing A Decentralized, Market-Driven Extension System in India: The ATMA Model. In *Changing Roles of Agricultural Extension in Asian Nations*. eds., A W Van den Ban and R. K. Samanta. Delhi. B.R. Publishing Corporation: 203-228.

* Spencer, Dunstan S.C. (1976). *African Women in Agricultural Development: A Case Study in Sierra Leone.*, Washington, D.C.: Overseas Liaison Committee, American Council of Education.

Sulaiman, Rasheed V., and Andy Hall. (2002). "Beyond Technology Dissemination: Can Indian Agricultural Extension Re-invent Itself?" New Delhi: National Centre for Agricultural Economics and Policy Research, Policy Brief 16. Available on line at *http://www.ncap.res.in/upload_files/policy_brief/pb16.pdf*

Sulaiman, Rasheed V. and Andy J. Hall. (2006). "Extension Policy Analysis in Asian Nations". In *Changing Roles of Agricultural Extension in Asian Nations*. Editors A.W. van den Ban and R.K. Samanta. Delhi: B.R. Publishing Corporation. pp. 23-54.

Sutherland, Alistair. (1988). "Extension and Farming Systems Research in Zambia". In *Training and Visit Extension In Practice*. Agricultural Administration Unit Occasional Paper 8:49-78. edited by John Howell. London: Overseas Development Institute.

Suttmeier, Richard P., Cong Cao, and Denis Fred Simon. (2006). "China's Innovation Challenges and the Remaking of the Chinese Academy of Sciences". *Innovations*. Summer: 78-97.

* Swaminathan, M.S. (1982). "Biotechnology Research and Third World Agriculture," *Science*. Vol. 218, 3.

* Swanson, Burton. (2006). "The Changing Role of Agricultural Extension in a Global Economy". *Journal of International Agricultural and Extension Education* 11(3): 5-17.

Swanson, Burton *et al.,*. (1997). *Improving Agricultural Extension: A Reference Manual*. Rome: FAO.

Torero, Maximo and Joachim von Braun. (Eds.) (2006). *Information and Communication Technologies for Development and Poverty Reduction: The Potential of Telecommunications*. John Hopkins University Press. IFPRI.

Tripp, Robert. (2006). "Is low external input technology contributing to sustainable agricultural development?" *Natural Resource Perspectives* 102:1-4. London: ODI.

* Tripp, R., M. Wijeratne and V. H. Piyadasa. (2005). "What should we expect from farmer field schools? A Sri Lanka case study". *World Development* 33(10): 1705-1720.

Tripp, Robert, *et al.,* (2006). *Self-Sufficient Agriculture: Labor and Knowledge in Small-Scale Farming.* London: Overseas Development Council.

Tripp, Robert, *et al.,* (eds.) (2006). "After School: The Outcome of Farmer Field Schools in Southern Sri Lanka". In *Self-Sufficient Agriculture, Labor and Knowledge in Small Scale Farming.* London: Earthscan.

Van de Fliert, Elske. (2006). "The Role of the Farmer Field School in the Transition to Sustainable Agriculture". In *Changing Roles of Agricultural Extension in Asian Nations.* (Editors) A.W. Van den Ban and R.K. Samanta. Delhi: B.R. Publishing Corporation. pp. 325-342.

* Van den Ban, A.W. and R. K. Samanta (eds.) (2006). *Changing Roles of Agricultural Extension in Asian Nations.* Delhi: B.R. Publishing.

Van den Berg, Henk and Janice Jiggins. (2007). "Investing in Farmers – The Impacts of Farmers Field Schools in Relation to Integrated Pest Management". *World Development.* 35(4): 663-686.

World Bank. (2005a). *Uganda – Millennium Science Initiative Project. Vol. 1.* Washington, D.C.: World Bank.

World Bank. (2005b). *Capacity Building in Africa: An OED Evaluation of World Bank Support.* Washington, D.C.: World Bank Operation Evaluation Department.

World Bank. (2006a). "Project Appraisal Document on a Proposed Credit to the Government of Ethiopia for a Rural Capacity Building Project". Washington, D.C.: May 18. Draft.

World Bank. (2006b). "National Agricultural Innovation Project, Republic of India". Report No. 34908-IN. World Bank. India Country Management Unit, New Delhi; India.

World Bank. 2007a. *Enhancing Agricultural Innovation. How to Go Beyond the Strengthening of Research Systems.* Washington, D.C.: World Bank.

World Bank. (2007b). *Revising Sri Lanka's Agricultural Research and Extension System: Towards More Innovation and Market Orientation.* May. Colombo: World Bank.

World Bank. (Forthcoming). *World Development Report 2008: Agriculture for Development.* Washington, D.C.: World Bank.

9

Agricultural Transformation in Latin America

Basistha Chatterjee

The wave of global market demand for value added and newly derived agricultural produce has touched Latin America to a great extent. Latin America has also the big supply potential especially the countries like Brazil, Chile, Peru and Guatemala. This paper envisaged on the challenges on Agricultural Transformation in Latin America which is led by global market force. But the transformation of agriculture is not homogeneous in Latin America. Only the skilled farmers and the modernized processing sector are enjoying the fruits of transformation where has the others are still stuck with traditional farming. Indeed, newly developed seeds, fertilizers and plant protection chemicals along with new technologies of modernized agriculture facilitated this transformation but until the small farmers can enter into the arena it cannot be said to be complete and healthy because small farmers encompass the big scenario in the Agriculture of Latin America. It is also a challenge because of the fact that, in spite of agricultural transformation the

level of poverty and unemployment remained static during the last ten years. The paper also suggests the possible solutions of the problem.

Introduction

Agricultural transformation is an innovative as well as integrated approach which rightly aims at i) To increase productivity and efficiency at all stages of the commodity chains, for greater value addition and employment generation, and ii) To reduce the cost of transaction between different stages of agribusiness system. Latin America is the fourth largest continent in the world covering an area of 17,821,018 square kilometers. The continent has the Caribbean Sea at its north end and Cape Horn is the South end which reflects its extended length. This continent is also famous for having the largest rainforest cover area in the world as well as largest mid-altitude grassland area in Pampas of Argentina. The agriculture of Latin America is definitely a transformed one which is a revolutionarily changed form of modernized one from a traditional one based on 'comparative' as well as 'competitive' advantage which is beyond national and continental boundaries in this era of globalization when the world is ever-shrinking. Moreover, at all stages of the commodity chain it has attained the economies of scale. Not only that, a vertical coordination is observed among the productive sectors and the service sectors.

The agricultural research, market of agricultural inputs, production level of the farm, agri-processing, adequate storage, well structured marketing system are the big factors of increasing the productivity and efficiency to the physical transformation and transformation of the commodity chains.

Transformation is not only the effort of improving an existing system but also is a process. This can be measured by its sustainability and stability. The first criterion is that the transformation process is molded by the existence of a diverse value in the society which clash, conflict and evolve into something new but does not suppress the other value system.

The second criterion is that the transforming society learns from the experience of others, assimilates innovation through a process on rational selection geared to stabilizing higher production and productivity of agricultural produce.

Large proportion of the GDP is provided by the agribusiness and food service sector which exhibits a healthy and promising forward linkage. Central America and Paraguay are agro-based countries while Brazil is an urbanized country but dependent on agriculture for growth. It is inevitable that in all the Latin American countries agricultural growth is important for reducing poverty.

The domestic agri-market of Latin America is highly transformed by the demand pull of the supermarkets. In this situation the heat of the challenges of global competition is felt by the smallholder farmers. Higher skill is required from non-farm economy where the access of rural poor particularly women is limited. The overall goal of the Latin American countries as a whole is to develop for promoting the smallholders in new food markets so that they may get remunerative jobs in Agriculture and the rural non-farm economy also sustaining the economy of the rural area.

Relationship between Agriculture and Livelihood in Latin America

It has been noticed that a high percentage of livelihood in the rural area of Latin America are directly or indirectly dependent on agriculture. It varies from country to country. As for example in Peru 35 per cent of the labour force is directly engaged in agriculture, whereas in Ecuador 7 per cent of the labour force is engaged directly in agriculture. Apart from that agriculture linked non-farm urban employment is also high. The linkage between agriculture and other economic sectors is very important for economic growth of any country individually and for the continent collectively. A strong linkage is the inter-sector growth multiplier will also be strengthened. It is also observed that the portion of budget which the urban sector of Latin America spends on food staples such as rice, wheat, maize, sorghum, soyabean, cassava, plantain and other fruits is fairly increasing which indicates the necessity of agriculture to the livelihood in this continent. Hence, the transformation of agriculture in Latin America from traditional to the modernized form was required for the following reasons:

a) Reducing poverty through broad-based economic growth.

b) Enhancing the security for food to approach towards self-sufficiency.

c) Creation of value added agricultural products and their marketing by judging market prospect and implementation of marketing plan with respect to that.

d) Export of the value added produces from the surplus produced countries and earn foreign currency.

A Bird's Eye View of the Transformation Status of the Latin American Countries

Although transformed agriculture is the overall picture of Latin America but the degree of transformation is varied within the ten countries namely Brazil, Argentina, Bolivia, Chile, Colombia, Ecuador, Paraguay, Peru, Uruguay and Venezuela. The present section will show the present status of transformed agriculture in these countries. The challenge on environment and agriculture will also be discussed.

Poverty and inequality is still persisting at a high note. In spite of 30 per cent increase in agriculture value added the rural poverty rate is unchanged in last ten years. Farmers are still feeling that the subsistence agriculture is the supporting cushion for them.

Brazil

Brazil is world's largest producer and exporter of coffee and orange juice concentrates and hence meeting the global demand of instant coffee in the supply chain with an increasing pace. Moreover, this country is the second largest exporter of soyabeans which shows the potential of this country. Not only that, this country has potentiality of high production and export of rice, sugarcane, cocoa and beet. As far as self-sufficiency of food is concerned, this country has attained the self-sufficiency in all food produce except wheat.

Despite deforestation of about 2 million hectares each year between 1979 and 1990, almost 58 per cent (1993) of Brazil remained forested. The rate of deforestation has also declined which is another important point to note because about 20 per cent of the identified plants are found here. Acid rain is a big problem here in agriculture.

Argentina

This country is one of the top five grain exporters in the world. Argentina is a densely populated country; most of the population lives in urban areas. The per capita agricultural production has declined between 1980 and 1997. The principal crops produced are wheat, maize, sorghum, soyabean and sugar beets. Agriculture contributes about 15 per cent of the Gross Domestic Production (GDP). The most potential resource is the fertile plain of Pampas which is needed to be utilized optimally.

Bolivia

This country has already attained the self-sufficiency in food. Agriculture accounted about 21 per cent of gross domestic production (GDP) of this country. The principal crops produced are coffee, cocoa, cotton, maize, sugarcane, rice, potatoes and timber.

Chile

Chile is one of the major fruit exporting countries in the world. Not only that this country has also attained self-sufficiency in most of the food crops. Including fishery and forestry agriculture accounts for 9 per cent of the GDP. The major crops which also have economic potentiality are wheat, maize, grapes, beans, sugar beets, potatoes and deciduous fruits. Obviously this country is also a major importer of food grains from the world market. Water pollution from raw sewage is the problematic issue threatening agriculture. This is an upcoming challenge for the future. However, the silver line is that more than 1 per cent of forest land is increased within 1980 to 1990.

Colombia

This country is one of the richest repositories of biodiversity in the world. The major crops produced are coffee, rice, tobacco, cassava (roots are eaten as staple), maize, sugarcane, cocoa bean, oilseeds, vegetables and forest product. Nowadays shrimp culture is gaining importance in this country. Agriculture accounts for 16 per cent of Gross Domestic Product (GDP) which is interesting. It is also to be noted that these crops comprised about two third of the agricultural output.

Ecuador

This country is the leading producer of Bananas and Balsa wood in the world. It is also a renowned exporter of coffee, cocoa, fish and shrimps to the world market. This country also has existing good productive potentiality of rice, potato, cassava, plantains (this crop appears like banana plant but eaten as staple) and sugarcane. Ecuador also imports foodgrains, sugar and dairy products. Soil erosion and flooding due to deforestation is the serious problem threatening agriculture. About 30 per cent of forest land has been lost from 1980 to 1992. However, rapid rate of deforestation declined in the last decade.

Paraguay

This country has attained it's self-sufficiency in foodgrains. 27 per cent of the Gross Domestic Production (GDP) is accounted for by agriculture. The cash crops produced are mainly cotton, oilseed and sugarcane. The other crops are gaining importances which are maize, wheat, yerba, tobacco, soyabeans, cassava, fruits and vegetables. This is primarily an agricultural country and economy is dependent on agricultures. The sewage and pollution are threatening problem to the livelihood as well as agriculture. This is due to the fact that massive forest destruction was occurred for agricultural purposes from 1980 to 2008.

Peru

This country has not yet attained the self-sufficiency in producing grain and vegetable oil.

The commercial crops produced in this country are coffee, cotton and sugarcane. The other crops which are potentially produced are rice, wheat, maize, beans, potatoes, plantain, cassava, fruits and barley. Agriculture of this country accounts for about 7 per cent of the GDP and about 35 per cent of is utilized in agriculture. An effort for conservation of endangered species is required urgently.

Uruguay

Uruguay has already attained self-sufficiency in foodgrain production. About 80 per cent of total land is used for agricultural and livestock grazing purposes. The major crops which are potentially produced are wheat, rice, maize, barley, sorghum, sugarcane, sugar beets and potatoes. Water pollution and acid rain are the problems which are major obstacles in the way of further development of agriculture. As a consequence of increasing pressure on agriculture, natural habitat of some species has been endangered. The instance of Eastern Wasteland is a burning example of the same. Deforestation led to soil erosion causing problem to agriculture.

Venezuela

About 2600 square kilometer of forest land was destroyed annually during 1970-1980. The agricultural produces are maize, sugarcane, sorghum, coconuts, rice, bananas, vegetables, cassava, coffee, fruit crops and cotton. Agriculture accounts for 6 per cent of the Gross Domestic Production (GDP). Though this country has already attained self-sufficiency in meat production, the self-sufficiency in food grain production is not achieved yet. Water pollution is a problem to be stated. During 1970-1980 about 2600 sq km of forest area was destroyed, which adversely affected agriculture of this country previously.

High degree of agricultural transformation took place in Latin America in an overall observation. About 75 per cent of agricultural produce is used up in domestic consumption. Supermarket channel is absorbing 60 per cent of the domestic retail sale. Export market is getting now offer of differentiated product with a fair trade such as organic coffee. As a consequence of these facts the farmers are also moving to specialized farming for vegetable production and organic cultivation.

Rural non-farm sector economy is diversified in Latin America. Agricultural labour market and the rural non-farm economy provides 70 per cent of rural income and 55 per cent of the rural labour force are employed in it. The need of non-farm operational skill, which is globally required is still away from the requirement due to market inequalities of finance and of market intelligence.

Constraints and Concerned Challenges for the Structural Transformation of the Agriculture in Latin America

Poverty remained the same throughout the last ten years in spite of the economic growth which is the prime area of concern. This reason for the same is manifold.

The priority given to agriculture was inconsistent, as a result, homogeneous progress did not take place. The causes for the same are formulation and implementation of the isolated projects of the crops, credit, irrigation, extensions, etc. The rural development programmes ware not integrated enough as per requirement. The modern structural adjustment and priority sector wise reforms also proved to be scanty. Among the other constraints the weak regional integration of the community chains, poor access to financing and insufficient investments, poor method of production as well as performance and market infrastructure, inadequate management of natural resource and poor tie ups between agricultural innovation and agricultural transformation are important.

Many countries of Latin America are also finding the way to reduce poverty by increasing agricultural income and rural non-farm economy as opposed to social assistance by subsidizing agriculture. It is being implemented by including small farmers in new food market by transferring their agricultural pattern from traditional low growth and low jobs to modernized high growth ad higher jobs.

Investment is done in agricultural sector for improving livelihood in subsistence agriculture and social assistance as education, skill training and health.

Emphasis by many countries local agricultural production system is given which can capitalize on the comparative advantage of that region so that new source of income is created. Priority is also given in governance for agricultural and rural areas. Market regulation from the ministry level is also an important need in this respect.

In this context it is inevitable to say that the fragmented market structure still exists in Latin America in National or sub-regional segments of sub-optimal scale to ensure profitability of modern private business environment. Moreover, with the increased import system the gap between various domestic production and increasing demand is met, from other continents as the fragmented market systems are increasingly open to global trade. This is a serious drawback in the pathway of successful transformation of Agriculture where the attention is needed.

But there are immense opportunities. The striking point in this regard is the natural potentiality of the Latin American countries to meet the global demand in a fashion of diversified harmony. For example, Brazil is the number one producer of highest quality coffee and orange juice concentrate, Argentina has the grain producing strength, Ecuador's banana producing capacity is world famous. Coffee and cocoa is a quite common potentiality of Latin America except for a few countries. This is a natural blessing which has to be utilized optimally which indicates the silver line of the future.

It is observed that other than inter-country exports within Latin America, major exports to USA, Japan, Mexico, Italy and Dominical Republic.

The 'issue plan' of agricultural transformation should be for the poor smallholder farmer who is the resource and should take into account the farmers' current practices and identify successes and shortcomings. It is necessary to determine and understand in as much detail as possible all relevant aspects of soils, vegetations, climate, fauna, pests, diseases. On that basis, the device of agricultural technique, their practices that are in harmony with the extent and nature of the country's natural resources which is eco-friendly and sustainable must be implemented.

So, for agricultural and rural development, establishing a target based production system is vital to ensure the acquisition of available technology and implementation of agreed development measures. Farmers have not been able to adopt research results because they have not been produced enough surplus where the currency is yield per hectare. The surplus yield is required to enable him to adopt cash as well as noncash input technologies. The only known strategy is to produce stabilized surplus to transform agriculture to sustainable and pull the resource poor small farmers out of poverty into the national economic stream.

The growing national and state consensus for upgrade quality of life, the rural-urban balance is required which is urgent. The change in the quality of growth is needed to make an effort to eradicate the poverty existing in Latin America. To achieve this, focus should be on improving the competition of small holding farmers' level in modern food markets, improvement of lifestyle through sustainable cultivation practices and improvement in the level of employment in farm and non-farm activities which is economically feasible. Government intervention on agricultural perspective in both national and local level is highly needed. In Latin American countries agricultural growth is 7 per cent from 1993 to 2005 which is not a very rosy picture. But some sparking growth has been observed in some sub-sector scenario. Soyabean in southern cone countries, biofuels in Brazil, fruits and salman in Chile, vegetables in Guatemala and Peru, cutflowers in Columbia and Ecuador.

Finally we dwell on the upcoming challenges. All the stakeholders i.e., Government, regional organizations, progressive farmers, agribusiness functionaries, development partners need to sustain momentum for agricultural transformation. Action plans are to be taken on priority basis and implementation of the same is required. Other than that, the land and water resource development, rural infrastructure and trade capacities for market access, food supply chain and responses to emergency food crisis (though majority of Latin American countries are now approaching to attain self sufficiency), agricultural research, improved agro-technology, dessimination and adoption of the same is requried.

To face these challenges the respective governments of the Latin American countries should create production-friendly environment followed by supervised subsidies preferably on a declining scale over a very short term of time frame. Also market strengthening policies should be incorporated after proper survey of the market. Setting up regulated markets with nearby warehousing capacity is very vital. Encouragement of co-operative formation by the government is essential in this respect; moreover, the setting up of crop wise specialized board is of utmost importance to develop the production quality as well as exploring new marketing potentiality. Agro export zone oriented export promotional approach is also required to be encouraged irrespective of government structures can develop every facet of agriculture in Latin America.

Conclusion

The continent Lain America had successfully fulfilled the criteria of agricultural transformation. The rise of efficient agribusiness system following the global demand is a wonderful fact which actually is the prime cause of this transformation. While making a comparison with the other continents Latin America is definitely far ahead of the other continents like Asia and Africa which shows the potentiality. Not only that the state of art of agriculture is dynamic which is interesting. The problems of poverty as a whole can be solved by appropriate policies by the policymakers. The immense potentiality of each and every corner of Latin America has to be explored and utilized optimally by maintaining the eco-sustainability.

(Dr. Basistha Chatterjee is a Faculty Member at Rural Management and Business Economics in IBRAD School Management and Sustainable Development, West Bengal, India. The author can be reached at basisthachatterjee@yahoo.com).

References

Agriculture for Development, "The Agenda for Latin America & Caribbean", World Development Report, 2008.

Beckford, G. L., Transforming Traditional Agriculture, Comment *Journal of Farm Economics*, November.1966.

Bichalo & Hoefle, "Divergent Trends in Brazilian Rural Transformation: Capitalized Agriculture in the Agreste & Sertao of the Northeast", *Bulletin of Latin American Research*, Vol.9, No.1, 1990, pp.49-77.

Boserup, E., The Conditions of Agricultural Growth, Aldine Publishing Company, Chicago, 1965.

Carstensen, V. "The land of Plenty" in Issues & Themes, American Association for State & Local History (through U S I S) New Delhi, 1975.

Dunn, J.M., "Review of Transforming Traditional Agriculture", *Journal of Agriculture Economics*, June,1965.

Goodwin, H. G., *Economics of agriculture,* Reston, Reston Publishing Co, 1977.

Google World Atlas- 2007.

Halcrow, H.G. *Economics of Agriculture,* Mc. Graw Hill International Book Agency.

Hazell, "Transformation in agriculture & their implications for rural development", *Centre for Environmental Policy*, Vol.4 (1), 2007. pp 47-65.

Kuznet, S, *Economic Growth & Structure*, Oxford & IBH Publishing Co., New Delhi. 1965.

Meheran, G.L.(1951), "Comparative Costs of Agriculture Price Support in 1949" Proceedings, *American Economic Review* May,1951.

Mellor, J.W. and Stevens, R. D. "The Average & Marginal Product of Farm; Labour in Under developed Economics", *Journal of Farm Economics*, August,1956.

Mellor, J.W., "Towards A Theory of Agricultural Development", Agricultural Development & Economic Growth, London, Cornell University Press 1953.

Metcaff, D, *Economics of Agriculture*, Harmondworth Penguin Books.

Nicholls, W. H., "The place of Agriculture in Economics Development", Agriculture in Economic Development, Vora & Co. Publishers Pvt. Ltd., Bombay, 1964.

Rosentein- Rodon, P. N. "Disguised Unemployment & Underemployment in Agriculture", *Monthly Bulletin of Agriculture & Statistics*, July- August, 1957.

Schultz, T. W., *Transforming Traditional Agriculture*, New Haven, Yale University Press, 1964.

Schultz, T.W., *The Economic Organization of Agriculture*, Bombay, TMM 1953.

Schultz, T.W. *Agriculture in Unstable Economy*, New York, Mc.Graw Hill, 1945.

Schultz, T.W., *Production & Welfare in Agriculture*, New York, The Macmillan Co. 1949.

Vlasin, New Economic Roles for the Rural Environment, Some key issues & challenges posed by non-agricultural Demands for rural development, proceeding paper of winter meeting of the American Agricultural Economic Association with Allied Social Science Associations, Detroit, December 27-29, 1970.

Witt, E, "Rode of Agriculture in Economic Development, A Review", *Journal of farm Economics*, February, 1965.

10

Agricultural Transformation
The Case of China

Asis Kumar Pain

The agricultural transformation in China was initiated in 1978 after almost three decades of failure to enhance the living standards of the rural population. The prime aspect of the transformation characterizes the changeover from a government controlled centrally-planned system to a market-driven one. As a result of the transformation, the agricultural output got doubled within a decade. Besides, there occurred substantial increase as well as diversification of the consumption pattern of the average rural Chinese. The manifestation of the enhanced consumption basket got manifested in the changeover from table grains and seasonal vegetables to all season's vegetables with high protein and calorific values, various meat and meat products along with a wide range of fruits for the average Chinese households. To understand the change, the phases, impact and the prospects of sustenance of the Chinese agriculture is examined.

Introduction

Agricultural transformation denotes a process by which production of individual farms get transformed towards catering a market, which is free of any subsistence and dealing with specialized products. To ensure it in practice, focus on the input and output delivery besides intimately integrating the operation with other sectors of the economy, both domestic as well as international are deemed essential. Thus, agricultural transformations automatically ensure transformation of the economy where apart from agriculture; the other sectors are involved in the generation of economic output and employment. Having discussed the theoretical underpinnings of the agricultural transformation, the practical side also needs exploration. The agricultural transformation in China is considered a bright example for such an illustration as after almost three decades of failure to enhance the living standards of the rural population, China resorted to reform of its agricultural sector in 1978. The central feature of such a transformation is the changeover from a centrally-planned system to a market-driven one. A remarkable feature of such a transformation is the doubling of agricultural output within a decadal span. As a result of such expansion, there occurred substantial increase as well as diversification of the consumption pattern. The food basket in the average Chinese households got supplemented from table grains and seasonal vegetables to all season's vegetables with high protein and calorific values, various meat and meat products along with a wide range of fruits.

In this backdrop, the present article tries to examine the different phases of the agricultural transformation in China as also, its prime impacts and sustenance in the future.

Agricultural Activities

It is a common fact that the geographical area earmarked for India far exceeded that of China. Estimates showed that the average availability of cultivable land per every rural Chinese household stands at three-fifth of their Indian counterpart, as on 1950. As a result of deficient cultivable land, multiple cropping coupled with increase of labour days per unit area under actual cultivation is practiced to ensure increased productivity. As a result of these practices, agricultural productivity was found to be quite on the higher side in China. But to further enhance productivity, so as to enable sustenance of the agricultural supply in line with the increasing demand, the requirement is to enhance the productivity further. The main constraint in that objective arises in the aspects of meeting the requirement of increasing investments as well as in the reclamation of land. To help in this effort, pooling the limited resources of the farmers was garnered. So as to counter the de-forestation momentum leading to soil erosion and degradation of the land, pooling of households was resorted towards ensuring collective effort for cleaning up investment. These above objectives specify the prime features of the agricultural transformation of China.

Agricultural Transformation in China – The Process

Agricultural Policy Reform (1979-93)

Prior to 1979, the agricultural policy pursued by China was aimed towards ensuring rural equity besides guaranteeing the provision of cheap food, labor and capital so that the country's main thrust of developing an industrial base does not get hampered. Towards that objective, state control of production and its marketing along with its trading was implemented. However, in that process, government usually lowered the price of procurement much below that of international ones. However, with the aforementioned strategies, the future of Chinese agriculture was found to be quite unattractive forcing the government to institute new alternative strategy in 1979 towards rejuvenation of the agricultural sector. As part of the new strategy, the procurement prices of agricultural produce were raised besides allowing rural markets as a place where farmers can sell their produce directly. Thereafter, in 1981, decentralization of agricultural production was resorted upon making room for farm households instead of a commune system. Within 8 years, almost the entire production comprised of the "Household Production Responsibility System." Further, the mandatory procedure of state procurement of agricultural commodities is reduced in terms of the number of commodities.

In the next phase of agricultural reform, the stress given was on liberalization of prices as well as introducing new strategies of marketing. In 1984, when there had been a bumper crop yield, the government eased the procurement restrictions further by introducing voluntary contractual agreements with the farmers. Thereafter, the grain market was allowed freedom from the shackles of four-decade-old grain rationing system in 1993. As a result of these measures, the selling price of almost the entire (approximately about 90 percent) agricultural produce of China was ascertained at their market-determined price.

The New Agricultural Policy (1994-the New Millennium)

As the sustenance of the agricultural reform process was at stake on account of the inflationary trend envisaged by the Chinese economy, the agricultural transformation under-way resorted to attractive marketing strategies. Rising grain prices was the prime force behind such an impediment causing increase in imports. To reverse the trend, the Chinese government resorted to provisioning of incentives in the form of higher procurement prices (above the world prices) with respect to certain types of crops so as to allure farmers to shift their agricultural production from oil seeds and cotton. As a result of the stance, in 1997, when China recorded the highest grain production, the basket comprised the smallest proportion of wheat (since 1961). The majority of the grain comprised of rice. Though these stances led the agricultural sector to substantially expand its sustenance, but it caused the government a substantial financial burden that in turn, led to deficiency of social welfare.

In 1995, as part of the agricultural reform process, introduction of "Governor's Grain Bag Responsibility System" was made whereby the provincial governors were held responsible for equilibrating the demand and supply of grain and in the process ensuring stability of grain prices. Such a measure was found to be quite successful on account of prices getting stabilized as well as getting decreased. However, as a result of implementation of such policies, there occurred sharp deviation in the price of grains in different provinces. While grain producers based in the coastal regions of China got the advantages of high subsidy, those in the other regions, most of whom are poor though with surplus yield was taxed much higher, leading to aggregate reduction in social welfare.

Also in 1997, a new policy was introduced and termed the "Four Separations and One Perfection" to take care of the financial problems of the government controlled state grain bureau as well as to take care of the government grain bureaus. The distinguishing features of the new policy are as follows:

a) The functioning of the policy of the bureau is separated from the commercial ones.

b) The reserve stocks of the government are separated from commercial purposes.

c) Secluding the responsibilities of central government from the local governments.

d) Distinguishing old bank debts from the new ones.

By ensuring the afore-mentioned separation, the "perfectionist" stance of the policy got embodied.

Again in 1998, the responsibility for the management of grain was vested with the provincial governments though the central government was vested with the responsibility of procuring the grains. The need for such a policy was to ensure stability of the price besides improvement in the marketing of grain by impeding competition. The same policies of slackening government control in the face of market liberalization were carried out thereafter.

Agricultural Transformation in China – The Groundwork

In order to understand the groundwork for the agricultural transformation in China, pre-reform official data were examined. It showed that in the pre-reform period from 1950 to 1959, the per-capita rural employment (in man days) increased from 119 to 189, with the major increase happening after 1955 when a shift to advanced cooperatives occurred. (Table 1). Such an increase in labour

Table 1: Annual Number of Days Employed Per Person in Rural China Over Years

Year/Days	Average Days (billions)	Total Annual
1950	119.0	26.489
1951	119.0	26.835
1952	119.0	27.168
1953	119.0	27.537
1954	119.0	28.155
1955	121.0	29.439
1956	149.0	38.084
1957	159.5	41.518
1958	174.6	47.474
1959	189.0	58.420

Source: China: Managing an Agricultural Transformation – Grain Sector Review; http://www.worldbank.org.cn/English/content/242y6112663.shtml

can be attributed to increased capital formation characterized as costless in nature. Further, as a result of the trend, the rate of investment rose sharply in the economy though its proper estimate was not made on account of the existence of the non-monetary part of such investment. Another distinguishing feature of such investment is its rise without adversely affecting the consumption pattern of the vulnerable population segment as a result of the egalitarian system of distribution followed. This egalitarianism, in the initial phases of the agricultural reform process, (upto 1978-79) despite leading to slow rate of increase in consumption of the vulnerable section of the population, (so as to ensure a trade-off with the ongoing industrialization drive) ensured a stable as well as high rate of consumption in the near future.

Looking at the negative aspects, agricultural transformation was effectuated at a juncture when the country was in the throes of a poor harvest on account of widespread flooding and pest attack in various parts of the country. Besides, between the period 1959 and 1963, there had been substantial downward movement in output causing problems for the effective effectuation of the transformation. However, as said earlier, the egalitarian system of distribution came to the rescue and spreaded the deficiency amongst a large section of the population without restricting the impact on specific segments. Thus, cases of able-bodied rural Chinese falling prey to malnutrition, disease and starvation remained very limited. Further, in the initial periods of reform, socialization of household activities, e.g., cooking and child minding provided the freedom of rural people especially the women who in turn, used their energy for agricultural productivity. However, a negative impact of the trend is social dislocation reflected in the postponement of child-birth causing drastic reduction in the birth rate.

Agricultural Credit – Situation and Source

Experience showed that accessibility of loans by Chinese farmers to be quite low as a result of scarcity of capital. Data revealed that in the Dongchen Village of Shaanxi Province in Northwestern China, only around 10 per cent of the entire 700 households have accessibility of agricultural loans. Also in the Sichuan, Guizhou, Yunnan provinces of Southwest China besides the Autonomous Region of Tibet, agricultural loans comprise a meagre 16 per cent of the total granted credit as on 2002. Examination of the trend showed that there occurred exodus of capital from rural to urban areas in the western and eastern areas as a result of retreatment of financial institutions from the rural areas due to various reasons. Of late, to resurrect the trend, the credit needs of the farmers are being sufficed by small and private financial institutions under the regulation of the government. The government on their part tries to allure investment, especially in the western part of China, which witnessed acute shortage of capital through higher rate of return. Besides, appropriate slackening of the restrictions for setting up of financial institutions in rural areas is being initiated.

Conclusion

As China undergoes transformation to achieve levels of development, the role of the government gets transformed from revenue generation through taxing to subsidizing the agricultural production. In line with this changing role policymakers had to make choice among various options. Amongst these, that of following the footmark of the Organisation for Economic Co-operation and Development (OECD) countries besides similar other newly developed economies, is considered a viable option as through such mechanism agriculture is protected through either appropriate support of price or through payments by direct farm income. However, in China, since agriculture provides employment opportunities for about 45 per cent of labour, such schemes becomes unviable on account of being too costly. Besides, providing price support could become detrimental for China during her entry into the World Trade Organization. Again, the other viable option could be to insulate the agricultural produce of China from international competition through development of appropriate trade barriers. However, in such a situation domestic demand and supply gets equilibrated leading to subsidization of the producers by the consumers. Though the option is alluring for a country like China as also congenial to her rapid economic growth but in such a setting, the chances of the agricultural market becoming distorted increases and with it decline of the economic efficiency. Further, like the first option, China's entry into World Trade Organization becomes uncertain. The next option is liberalization of the agricultural sector as quickly as possible. In such a situation, there arises no requirement of either taxing or subsidizing production with the agricultural produce to be exported or imported on the basis of China's comparative advantage scenario. The final option could be to liberalize her agriculture on a gradual basis. Such a step shall be in conformity with the World Trade Organization guidelines apart from enhancing the problem of food scarcity (with the rising population) which in turn, increases the economic efficiency.

Chinese policy-makers shall have to choose from amongst these options to increase the agricultural investment which had stagnated after China embraced agricultural transformation in 1978. Data showed in 1994, such investment was 32 percent below the 1978 level. Only after the investment in the agricultural sector gets stabilized, shall the possibility of ensuring a sustainable agriculture becomes bright.

(Asis Kumar Pain, Consulting Editor, Icfai Research Centre, Kolkata. He can be reached at asiskumarp@iupindia.org).

References

China: Managing an Agricultural Transformation – Grain Sector Review; ***http://www.worldbank.org.cn/English/content/242y6112663.shtml***

Shenggen Fan and Marc J. Cohen (May 1999); "Critical Choices for China's Agricultural Policy"; 2020 Brief No. 60; *http://www.ifpri.org/2020/BRIEFS/NUMBER60.HTM*

Utsa Patnaik "The Economic Ideas of Mao Zedong: Agricultural Transformation"; Socio-Political Changes and Economic Development.

Improving Farmers' Access to Bank Loans; *China Daily,* October 8, 2003.

11

Agricultural Transformation in India

Basistha Chatterjee and Visvarup Chakravarti

With the advent of the Green Revolution in India, agricultural transformation was rapidly visible in 1960-70, mainly boosted by high yielding variety seed of wheat and rice. Afterwards, the GDP growth continued steadily especially at 4 per cent during 1992 to 1996 and 2 per cent during 1997-2003. The share of rural poor and persons below poverty level has also reduced which is a noticeable feature. The subsidizing policy on fertilizers, and electricity for irrigation also facilitated the rapid transformation after 1970 but on the other hand pressure on the government budget also exerted simultaneously. This paper emphasizes on the consequences of this transformation viz., improved health, education, infrastructure and land use at an early stage. This normally varied from the plains (of the states of Punjab, Andhra Pradesh, Karnataka, Maharashtra) to the hilly tracts of Himalayas. Also the paper talks about the upcoming challenges – further prosperity, need for employment, limited natural resource.

Introduction

India has been making so many efforts to achieve self-sufficiency in food crops and to cut down the dependence on imports from other countries. Since Independence, India has taken up a policy of food sufficiency, especially in terms of food grains with the enthusiastic support from the Government. Agricultural policies were set by the Centre as well as the State Governments, which targeted at increasing crop production by expanding irrigation, improving yields by adopting high yielding varieties and increasing the intensity of cropping with the help of multiple cropping patterns. Now, more than 39 per cent of the area is irrigated. The overall agricultural production has increased at an average annual rate of 2.7 per cent. This is an important achievement as the population growth is increasing; the country has to keep pace with it in terms of the food production. As a result, India has now established herself as the world's leading producer of major crops including rice, wheat, coarse grains, pulses and cotton. Besides these, India is now one of the major milk producing countries. India also ranks high in the the fruit and vegetables production. As regards commercial plantation, India is one of the largest producers of tea, spices, cashewnuts, mangoes and bananas. The production of exportable horticultural products has also increased in the last few years.

In recent years the share of Agriculture in India's GDP has dropped which indicates a structural shift in the composition of GDP. Traditionally, this sector accounted for 40 per cent of GDP, but in recent times, a decline is observed in this respect. India's GDP fell down from 7.8 per cent in 1996-97 to 4.4 per cent in 2000-01. In order to increase the yields the Government is providing fertilizers, power and water for irrigation at subsidized rate to the farmers. Also a lot of public investments in Agricultural research, extension and infrastructural development facilities were provided. The national agricultural policy formulated by the Government of India fixes minimum support price for the major commodities, and update of prices is done each year for the same. The State Governments are empowered to provide better irrigation, power and fertilizers to the farmers. The Government gives the support price to the farmer and lifts their produce at times of natural calamities. Thus, the public stock disbursement among low-income consumers at subsidized prices is maintained through the Public Distribution System (PDS). On food subsidy India spends a huge portion of the budget. These efforts to increase the production as well as the price policies generated a steady growth in Agriculture ever since the Green Revolution in 1960s. The domestic market was insulated from the wave of global trade, with the restrictive policies imposed by the Government. High tariffs and on-tariff barriers narrowed the scope of market access. Now these restrictive policies are webbed and the total agricultural trade has been doubled between 1991 and 1999 which is an important proof of Agricultural transformation following the market-driven theory of transformation as stated by T.W.Schultz.

Three Phases of Agricultural Transformation in India

The effort of Government of India to transform Agriculture from its traditional way to modernized one were based on:

i) Inventing the new agricultural input which have a relatively high pay off and to make them available to the farmers.

ii) Supplying these inputs to the farmers and ensuring the process.

iii) Making the farmers learn about the process of how to use them efficiently.

If we go through the time frame wise study we can split the entire progress of Agriculture after independence in three phases according to the nature of transformation.

A) From 1947 to 1961

A lot of institutional and infrastructural changes were introduced during this period, with the objective to increase crop production by motivating the farmers. Major institutional changes happened during this phase, like the abolition of the zamindari system, tenancy reforms, ceiling on land reforms, ceiling on land holdings and consolidation of holdings. The infrastructural development such as construction of new roads, provision of additional irrigation facilities through multipurpose projects. The stress was given to provide required agricultural credit to the farmers and to produce more fertilizers. In this phase the Community Development Programme was started for the benifit of agriculture and rural development. During this period several Agricultural Universities and Agricultural Research stations were established. As the required inputs were lacking the desired level of Agricultural growth was not possible to attain. The Ford Foundation experts recommended concentrating on certain crops and certain potential areas. This phase was termed as the general phase. In this phase the foodgrain output was increased from 51 million tonnes in 1951-52 to 68 million tonnes in 1955-56 and 82 million tonnes in 1960-61.

B) From 1961 to 1965

To concentrate on the potentiality of the local resources and their proper utilization the Intensive Agricultural District Programme (IADP) was evolved in 1961. A package programme was taken such as improved seeds, improved implements, balanced use of fertilizers, manures and pesticides for all important crops. The technical know-how of the inputs were provided in selected districts followed by credit, soil testing and plant protection measures. As a consequence of which the food grain output increased to 89 million tonnes in 1964-65 as compared to 82 million tonnes in 1960-61. In 1964-65 a modified IADP programme called Intensive Agricultural Area Programme (IAAP) was introduced with specific crops. The support price policies were set up from this period; this phase was termed as Intensive Phase.

C) From 1965 to 1995

In this phase the higher yield of a particular crop with a higher cropping intensity was targeted. New agro-technology played an important part along with new crop rotations. Short duration varieties were evolved for rice, wheat, maize, jowar and bajra for specific agro climatic conditions. The concept and application of high yielding variety, multiple cropping, modern farming and developed irrigation are the features of this phase.

Also various rural development programmes were initiated to increase the receptivity of the farmers. The two drought prone years 1965-66 and 1966-67 were a setback for food grain production as well as the economy. A steady increase in food grain production was observed in 1967 to 1971. The increase in yield per hectare was also associated with diversion of land from other food crops to rice and wheat. This period was termed as the 'Green Revolution' by the economists. It is also interesting that after the Green Revolution the relative as well as absolute increase in the production of wheat was more than that of rice production. Rice production got its popularity in wheat producing areas of Punjab, Haryana and western part of Uttar Pradesh. The wheat cultivation gained popularity in the states of West Bengal, Maharashtra and Gujarat. As a result of which the regional imbalances and personal income imbalances were decreased to a great extent. This huge increase also proved that illiteracy is not the prime constraint of the transformation of traditional agriculture of India, because the farmers were able to use new technical inputs very effectively. Also the efficiency of the Indian farmers, technical inputs and extension program was proved. The Green Revolution happened because of the result of supply of new inputs of agriculture, multiple cropping programme, agricultural credit facilities, intensive prices of agricultural commodities, development programmes for small and marginal farmers by some previously formed rural development agencies like Small Farmers' Development Agency (SFDA) and Marginal Farmers' and Agricultural Labourers' Agency (MFALA), to make the farmers empowered to adopt the technology to increase production. The supply of new inputs of agriculture encompasses adoption of high yielding varieties of seeds by National Seeds Corporation, State Seeds Corporations and Agricultural Universities, supply of huge amounts of chemical fertilizers, which is required for the production of high yielding varieties of seeds, expansion of irrigation facilities especially the contribution of minor irrigation projects like wells, tube wells are quite significant in this aspect which reduced uncertainties of crop production, plant protection and pest controlling chemicals, development of infrastructures like transport and communication, regulated market, storage and water housing, agricultural education and power and use of machinery of improved quality e.g., tractor.

The interesting proof of agricultural transformation in this period is that the net area sown in 1951-52 was observed to be 11.87 crore hectares whereas in 1993-94 it was 14.21 crore hectares. Not only that, the multiple cropping area in 1951-52 was 1.32 hectares (11.10% of net area sown) and in 1993-94 was 4.43% (30% of net area sown).

The outcomes of the progress of this Green Revolution periods are manifold. The food grain output increased considerably which rose from 51.00 million tonnes in 1951-52 to 191.50 million tonnes in 1995-96. The sufficiency in food is indicated by the reduction of import of food grains which was 6.64 million tonnes in an average in the initial years, later it fell down to 1.83 million tonnes.

Trend of Food Grain Production During Last three Decades

Year	Production (million tones)
1970-71	108
1972-73	95
1978-79	132
1979-80	108
1990-91	176
2001-02	213
2002-03	174
2003-04	212
2006-07	216

Sources: 1. CMIE, Basic Statistics Relating to Indian Economy, Vol.I, August 1994.
2. Various Economic Surveys.

Trend of Food Grain Production in India

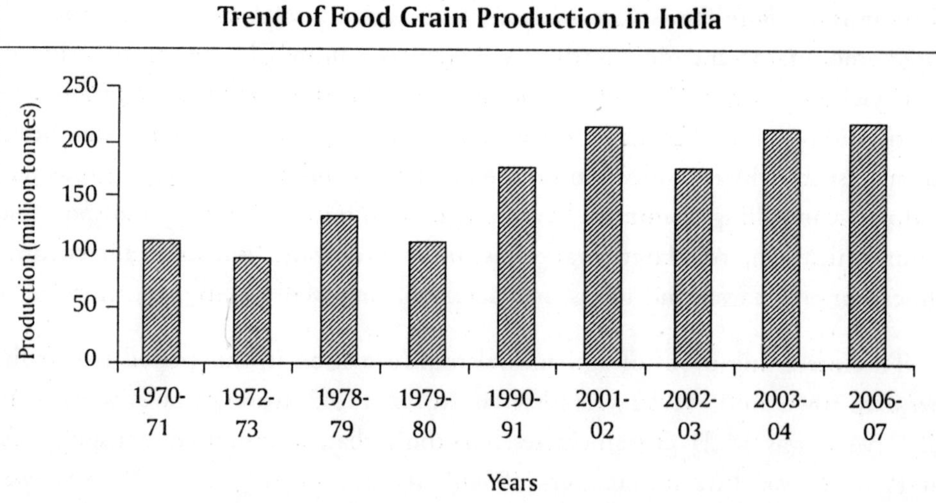

Sources. 1. CMIE, Basic Statistics Relating to Indian Economy, Vol.I, August 1994.
2. Various Economic Surveys.

The trend in the foodgrain production in India shows that after some fluctuations the output of food grains production rose to 176 million tonnes in 1990-91 and touched 213 million tonnes in 2001-2002. Due to drought the output declined sharply to 174 million tonnes during 2002-2003.

Augmentation of employment in farm and non-farm activities were generated as a result of multiple cropping system, development of agro-industries, marketing and transportation sectors.

The overall rural development was observed which is supported by several types of construction activities in rural areas. In this respect the establishment of banks in the rural areas is also a striking feature. The income of rural people also increased as the improvement of standard of living was noticed as a result of new technology.

While considering the benefits fetched by the Green Revolution it was also noticed that the states like Punjab, Haryana and west part of Uttar Pradesh reaped the benefits of Green Revolution which was not the case in the states like Rajasthan, Himachal Pradesh, Bihar and Assam. These disparities and unbalanced economic growth is the challenge for Indian Economy. There is also the disparity of income distribution, as the small and marginal farmers could not fetch the benefits of Green Revolution. The wide range of employment in the large farm sector was also discarded due to mechanized farming in the large farm and a capitalistic nature of farming was generated in large farm as supported by the report of All India Rural Credit Review Committee of the Reserve Bank in 1969. Other than that the high price of land, storage problem and increasing cost of production are also the constraints which were coming in the way of modernization of agriculture.

Transformation of Agriculture with Special Reference to Hilly Tract

Unlike the plains, the hill agriculture has never targeted to maximize production, but rather optimize the same and at the same time tried to maintain the quality through holistic and traditional agronomic practices and thus remains well behind the Gangetic plains and Deccan plateau from the view point of quantitative output. The small land holdings, rugged terrains, different edaphic situation, land and soil erosion, seasonal water sources and comparatively backward irrigation management – all these contribute to the crop selection and cropping pattern of the entire Himalayan tract.

Crop diversification is the main subject of interest in the Himalayan agriculture which is to a great extent, dependent on forest resources. Traditional Himalayan agriculture emphasizes crop-livestock mixed farming, which depends on fodder from forests and manure from forest leaf litter and farm animals excreta. The influence of forest has been very high on traditional agriculture because of various forest-driven tangible and intangible benefits (mainly NTFPs and services) that the hill people enjoy and are at the same time crucial for their livelihood. As a result of transformation the traditional staple food crops have been gradually replaced by the horticultural cash crops and fruit crops. The traditional varieties have almost been abandoned.

State wise scenario is discussed below:

Jammu and Kashmir

In the valley of Kashmir wherever soil is conducive for growth and the irrigation source is available, crops like rice, corn, millets, barley, wheat and potato are grow. Rice and wheat both are obviously cropped at low to middle altitude. Among spices, saffron is the most important crop for the middle altitude Kashmir valley. Most of the India's temperate zone fruits like apple, walnuts, apricot, cherries, etc., are grown as fruit crop in the valley as well as in the high altitudes of Himachal Pradesh and Uttarakhand. Rajma (*Dolicos lablab*) cultivation is highly preferred in Kashmir for its high demand in domestic market and export.

Many farmers are gradually taking up medicinal plants cultivation as crop diversification as the market for Medicinal and Aromatic Plants (MAPs) is developing, specially for high altitude MAPs. A few Growers cooperatives have been looking after the marketing as well as cultivation. Kashmiri Kuth (*Saussuea costus*), Pushkarmool (*Inula racemosa*), Sahjeera (*Carum carvi*) etc., are a few of the MAPs, gaining ground nowadays. Considering the overall agriculture scenario of the state following steps may be taken up:

a. It is now necessary for the state to stress the need and adopt the latest agriculture practices and technologies and reach the same to the farmers through different participatory modes of agriculture extension, under strict monitoring mechanism to have controlled implementation of various schemes at grassroots level in the state.

b. Industrial status should be given to agriculture and horticulture, which would enable growers and farming community to avail various incentives.

c. Proper and timely harvesting of the available water resources for irrigation purposes by adopting innovative methods avoiding available standard pesticides, high yielding seeds and chemical fertilizers and including organic fertilizers and pesticides.

d. Cultivation of traditional vegetables and judiciously off-season vegetables to minimize dependence on import of vegetables.

e. To industrialize mushroom and spawn production in the state especially in the Jammu division.

f. To introduce latest post-harvest handling technologies to improve the quality of fruit production, which is imperative to compete in international market.

g. Less use of synthetic fertilizers in J&K is an important issue for maintenance of quality of agriculture procedure and also for better land and soil management.

Himachal Pradesh

The western Himalayan state is famous for the vegetables cultivation. The competing needs should be balanced to increase yield by optimizing environmental development and income escalation. The high Himalayan regions of Himachal Pradesh consists of three districts of Chamba, Lahaul & Spiti and Kinnaur. A chunk of the said geographical area comes under rain-shadow region, which receives minimum rainfall but high snowfall and thus controls the agri-management of the area. In other regions rainfall is high and widespread. The agriculture is generally practiced in the river valleys. In the low altitude areas with good soil condition and good irrigation facilities rice, wheat, etc., are cultivated along with various vegetables like carrot, cauliflower, cabbage, chillies, raddish, beans, etc.; in the rain-shadow areas the main crops are barley, pea and soybean, which is a newly introduced crop in the area.

"Of the total landholdings 42% are less than 0.5 ha in size and 22% are 0.5 to 1.0 ha, which means 64% peasants are marginal farmers in Himachal Pradesh. The majority of them would be living in abject poverty if not supported by CPR and non-farm activities (Bhati and Zingel, 1996)". With the advent of alien crop introduction, increase of biotic pressure, indiscriminate use of toxic chemicals, coupled with loss of barren lands and associated biodiversity loss, poor regenerative capacity, soil erosion and poor carrying capacity, unsustainability in hill agriculture has been aggravating.

The overall agricultural scenario in Himachal can be summarized as follows:

a. Traditional crop races are being gradually replaced by high yielding varieties which are more vulnerable to diseases and cultivation has become more cost- and input-intensive;

b. Quality of soil has been falling down and at the same time erosion related problems are coming up;

c. Tomato, bell pepper, rajma, soybean, suitable high yielding varieties of cabbage, cauliflower crops, which could fetch more economic returns, attract farmers more and gradually they are taking up the same;

d. The traditional varieties are becoming less in use although there are few organizations which are in favor of the same;

e. Increased anthropogenic pressure is one main reason for biodiversity loss;

f. Fragmentation and loss of habitat also add to the causes of biodiversity loss;

g. Arable land quality and percentage both have decreased considerably;

h. Fruit crops like apple, apricot, walnuts have been gaining ground. The first one requires synthetic spray of pesticides, which affects soil and under-grown crops and that's why combination crop selection is also highly difficult; and

i. Potato cultivation is an attractive proposition; Potato Growers' Society is having a flourishing business.

In a few pockets of high altitudes hopps cultivation has been gaining ground due to its ready and easy industrial application. At the same time because of global as well as domestic market demand few farmers are coming up with MAPs cultivation.

The crops include *Saussurea costus, Inula racemosa, Aconitum heterophyllum, Picrorhiza kurrooa, Carum carvi, Bunium persicum, Valeriana jatamansi, Angelica glauca,* etc., which have a substantial market both in the domestic and international arena.

Uttarakhand

In Uttarkhand, the only Central Indian Himalayan State, the traditional agriculture is still *in vogue* in some pockets of the high altitude, while in some other areas especially in the tropical and sub-tropical areas the high technology-intensive agriculture has become *in vogue*. The agriculture is generally practiced in the river valleys of the Tons, Yamuna, Bhagirathi and Alaknanda the four major rivers of UA. In this context it is relevant to mention that the state has been declared as 'organic'. Therefore, less emphasis is given from State Government on inorganic fertilizers and in many areas transformation of land has been in progress. The dominant cultural practices of the plains also threatens Uttarakhand's traditional cropping pattern. For example, mill-polished rice has out-competed mountain crops. Dehradun rice has got a massive attack. The area which was once renowned for Basmati rice, reports very less area now for paddy cultivation. The concrete jungle has replaced the same. Most of the traditional varieties are now out of cultivation although some varieties have been restored through some NGO activities 'Save the Seeds movement' and raising awareness about the need for maintenance of agricultural biodiversity.

The land ownership in Uttarakhand is unique in India in the sense that most farmers are owner-cultivators. Tenant farming and share-cropping are rare, and landholdings are generally small and limited to family farm which is mostly marginal i.e., less than 1 ha. As such, the zamindari system of big landholders is limited to the plains. In the hilly tract geography and hilly (Parbatyo) cultural heritage dominated the maintenance of traditional equitable land distribution in the state.

In river valleys of Uttarakhand mainly in the valleys of Alakananda, Bhagirathi, Mandakini and Yamuna and Tons following types of agriculture systems are observed and. the practice depends on the category of land to which the system must suit.

In the Tarai land i.e., lower Dun wheat, rice and sugarcane are the main crops in the rain fed system while in lower Shivalik, wheat, rice, ragi (local name Mandua; sci. name *Elucine coracana*), Jhingora (*Oplismenus frumentaceus*), maize and chaulai (*Amaranthus polygamous, Amaranthus blitum*) are grown. Although the land is squeezing day by day and soil fertility is declining, one kind of Basmati rice 'Khaadar Basmati' is still famous in the upper Dun. In the unirrigated area mandua, jhingora and chaulai – all are grown and they are well established in terrain condition. All the three species are suited to give good economic returns. A rare spice has recently come under cultivation called 'Jakhia', which is suitable to grow in the tropical to sub-tropical areas of Uttarakhand. In the forest fringe lands also *Eleusine coracana, Oplismenus frumentaceus and Amaranthus polygamous, Amaranthus blitum* are the main crops. In low altitudes double cropping is being practiced with the crops like rice, wheat and sugarcane. In the temperate Garhwal and Kumaon, mandua, jhngora, rice, wheat, potato, barley and cheena (*Panicum miliaceum*) are important to mention. In the temperate Himalayas of the state apart from potato, barley, chaulai, kodo (*Fagopyum esculantum*), uva (*Hoycleum himalayanse*). In temperate Uttarakhand Rajma is one of the most remunerative crops. In addition, they also cultivate hill varieties of spinach, chestnut, mushroom, *Angelica glauca, Carum carvi Jambu*-faran (*Allium carolinianum, Allium stracyii)* etc. In comparatively lower altitude Masur, Kulath and various types of Rajma are cultivated. In Uttarkashi varieties of rajma are cultivated.

Among vegetables and spices pumpkins, beans, corn, ginger, chilli, cucumbers, leafy vegetables, and tobacco are also grown. Potatoes have become an important cash crop, growing in areas unsuitable for other plants. During early winter after the rainy season (millet) and midsummer before the hot dry season barley-wheat combination is preferred.

In the Tons valley the Kharif crops include *Oryza sativa* (Dry and wet varieties), *Zea mays* (in lower altitude HYVs are cultivated while in the high altitude traditional varieties are cultivated), *Amaranthus candutus, A. crutenus, Chenopodium album, Echinochola frumentacea, Elucine coracana, Aspalum scrobiculataum, Setaria italica, Sorghum vulgare, Paspalum scrobiculatum* and *Glycin max;* the last one is a newly introduced one. Other potential crops are *Vigna mungo, V. radiatea* and *V umbellata* available in wide range of varieties.

In the Rabi season *Triticum vulgare* (wheat), barley (*Hordeum vulgare*), *Pisum sativum* (Green pea), *Cicer arietinum, Lens culinaries* are important.

To cope with recent development of MAPs sector, many farmers have taken up medicinal plants cultivation in which *Saussurea costus, Aconitum heterophyllum, Aconitum violaceum, Picrorhiza kurrooa, Valeriana jatamansi, Nardostachys jatamansi, Angelica glauca* are of maximum importance. In middle-upper altitudes, geranium (*Palagonium graveolens*) has been well adapted and its volatile oils also efficiently extracted and marketed. Few people have also been trying with Damascan rose for its highly remunerative essential oil.

Sikkim and Arunachal Pradesh

Sikkim and Arunachal Pradesh constitute the eastern Himalayas. Sikkim Himalaya also shows rapid land cover changes coming under influence of heavy anthropogenic pressure. The difficult geography clubbed with habitat fragmentation and loss, soil erosion and heavy rainfall have made the situation very vulnerable and influences the pattern of land ownership, subsistence agricultural production and the market perspectives. The traditional agro-system is still being emphasized by the state Government foreseeing the negative impact of intensive agriculture and targeting to make the state an organic one. Therefore, the mixed cropping pattern is followed giving importance to both crops and cattle. Forest generated leaf litter and Farm yard manure both contribute to agri-production to a large extent while inorganic fertilizers are not given priority. Although the productivity is very low, horticulture and spices are the major emphasized crops.

The priority crops include rice, wheat, maize and potato. The major spices are large cardamom and ginger, the former is most important because the state heads in production of large cardamom. Orange is another horticultural crop which receives heavy importance as it is money earning. Off season vegetables are the other items, the state prefers to cultivate.

In Arunachal Pradesh agri-operations includes both Jhum system and permanent cultivation. The tribes like Adis, Akas, Apatanis, Bangnis, Nishis, Mishmis, Mijis, Tangsas and others are still practising jhum or shifting cultivation.

In the conducive edapho-climatic situation and topography rice, millets, wheat, pulses, sugarcane, potato are cultivated. For crop diversification Kiwi is another foreign fruit which is given priority in Arunachal and its market is growing. The floriculture is also growing as the state has treasure of orchids. Different national and international bodies have been working in this field and carry on research.

Recent advancement in the behaviour of the farmers shows they are gradually adopting MAP cultivation for better economic return. Experimentally found that Kashmiri Kuth (*Saussurea costus*) can grow well but its market is still under question. *Panax pseudo-ginseng* is another promising crop but its domestication and cultivation standardization is still awaited. National Medicinal Plant Board (NMPB) has been emphasizing the matter nowadays.

Recent Progress and Achievements of Indian Agriculture

India has made a lot of progress in agricultural aspects in recent years. As a result of that India is the largest producer of milk, fruits, cashew nuts, coconuts and tea in the world. Moreover, India is now on the second position in terms of producing wheat, vegetables, sugar and fish. Also this country holds the third position in the world for growing tobacco and rice. Nowadays India is transformed from a food deficient country to a food surplus country because of improvement in irrigation, fertilizers, quality seed, proper applications of fertilizers and agrichemicals (pesticides) and agricultural extensions. It has now a reserve of more than 60 million tonnes of food grains. The per capita availability of food grains is now 500 gms per day which was just 350 gm in 1951 despite the increase in population to 100 crore from 35 crore. The overall food grain production during 2001-02 is a record production of 211.17 million tonnes, which is 12.25 million tonnes higher than that of the production of the previous year. So far the highest production of food grains had been achieved in 1999-2000 which is 209.80 million tonnes. The allocation for agriculture was increased by 58 per cent through enhancement of agricultural irrigation and agricultural credit in the 1998-99 budget. The allocation of accelerated irrigation programme was increased by 200 crores. The share capital of National Bank for Agricultural and Rural Development (NABARD) was raised by 500 crore. An amount of Rs.264 crores was given for rehabilitation and recapitalization of the regional rural banks. Not only that the allotment of Rs.100 crores for experimental crop insurance scheme and Rs.500 for rural infrastructure development fund as a provision arranged by NABARD is an interesting point to note. The budget was increased recently for subsidy to potassium and phosphatic fertilizers. The reduction of import duties on bio-pesticides and excise duty of jute, tax on bio-pesticide manufacturing units and duty on diesel generating sets are the wise steps of the outlay of central plan of India.

The announcement of National Agriculture Policy in July 2000 is also an important step which seeks the untapped potentiality of Indian Agriculture, to strengthen rural infrastructure, promote value addition, accelerate the growth of agri-business, employment generation in the rural areas, augmenting a sustainable standard of living for the farmers and face the emerging needs of globalization.

For the rainfed areas National Watershed Development Project is going on which aims at treating 2.25 million hectares of land with a funding of Rs.200 crore in the ninth plan with a holistic as well as integrated system approach involving community participation and ownership. In the 171 districts of the states namely Uttar Pradesh, Bihar, Jharkhand, West Bengal, Assam, Orissa, Chhattisgarh, Manipur, Mizoram and Arunachal Pradesh the on farm water management in eastern India has been launched for sustainably increasing the yield of different crops especially in the eastern region of India. Government is providing bank loans and subsidies to implement the scheme through the bank.

To utilize the unemployed agricultural graduates to provide the extension services to the farmers as commercial consultancy basis as well as self employment the scheme of Agri-clinic and Agri business Centres has been launched in 2001-02.

The Kissan Credit Card is also an important feature which is recently launched in 1999-2000 to provide timely support by the banking system to the farmers for their cultivation purpose in a flexible and cost-effective manner. This scheme has shown a rapid progress and over 2.90 crore kissan credit card had been distributed already as on 2002.

At last it is the emerging focus of India to look at the future trading. Fiture trading of gur, potato, raw jute, turmeric, pepper and castorseed in going on. Just recently future trading of coffee, castor oil, jute sacking, cotton, palmolin oil and oilseeds are approved by the government subjected to fulfillment of necessary conditions. In this context the international future exchange in pepper which started at 1999 in Navi Mumbai is an important step.

Challenges and Concerns in the way of Agricultural Transformation

A gradual decline in the growth of agriculture and allied sector is observed presently due to increase of drought. This is affecting the GDP growth rate of the recent years. This is a concerning situation that the growth rate of food grain of India just matches the annual growth rate of the population and the population below the poverty line is 40 per cent. The production of food grain fell to 192.40 million tonnes in 1997-98 from 199.40 million tonnes in 1996-97. The year 1998-99 was not so good though in 2001-02 better achievement was there.

The large disparity in the country adversely affects the growth of agriculture in India – differences in climate, availability of infrastructure, irrigation, etc. The cause of the disparities are the uneven development of agriculture in different regions either because of differences in irrigation facilities or because of the suitability of regions for producing rice and wheat or due to the differences of other inputs which are complementary in nature. Also the difference of size of holding is a key issue in this context. A comparatively higher level of agricultural development in certain states; the characters of farmers and the administrative efficiency are crucial factors. Fortunately, the Central Government has realized the problem and food and price policy along with a central plan of action has been taken to meet those problems.

Suggested Ways to meet the Constraints

Specifically for the Plains

The study of the transformation of agriculture in plains reflects that the yield of the crops are still subjected to the weather conditions to a great extent though the maximum and minimum total output are much more than the past. Some regions got prosperity and growth of capitalistic farming took place whereas some other regions remained backward and underdeveloped. The following suggestions may be followed in this respect.

- On the basis of the current scenario of agriculture of India there is need for a region-specific policy in which prior consideration should be given to agro-climatic situation, suitable species and regional economic status.
- Mono-cultural practices should be replaced by suitable multi-cropping systems with the option of crop rotation. Where multi-cropping is not possible crop-rotation or mixed cropping opcions should be explored.
- The policy should aim at more agri-based industries that would generate more employment and at the same time it would be more market-oriented.
- The new policy should emphasize on crop diversification which would not only generate alternate livelihood but also restore soil fertility and augment/ optimize production.
- Domestication of wild varieties and development of their cultivation protocols should be prioritized under the sector of research & development.
- Limited use of chemical fertilizers and pesticides and more use of organic manure should get preference. That would result in sustainable production of healthy crops.
- Choice of suitable agro-forestry system according to the edapho-climatic situation, eco-floristic zones and forest types would not only result in judicial and economic use of land but also diversified production from the same field.
- Farmers' aid is another point of concern. Crop insurance should cover more number of crops and farmers all over India should get its benefit. Likewise for some specific agro-crops buy-back policy should be encouraged as the same would give better encouragement to the farmers to take up some crops which have fluctuating markets.

Specifically for Himalayan Hilly Tracts

The agriculture in the rough and difficult terrain of the Indian Himalayas have numerous challenges. The percent effective forest cover in the Himalayas, on one hand, is comparatively larger than the plains. In fact, the residual forests of the hilly tracts contribute a lion share to maintain the National Forests in a bit satisfactory condition. On the other hand, the forests of the hilly tracts are the repository of different highly significant traditional or folk varieties including edible, horticultural and medicinal and aromatic plants. If these forests are gone the agro-biodiversity of the region will be highly affected.

The climate change in the recent times, also looms large on biodiversity maintenance environmental conservation and agricultural production sustainability. Among other hazardous components come undulating topography, unavailability of large stretch of plain lands (the existing lands are generally marginal), soil erosion and excess run off. In cold desert regions in the high altitude Himalayas the erratic rainfall (less than 50 mm) and difficult water access pose problem in selection of crop for the region. The wetlands and natural lakes, which provide immense contribution to the regional ecology and agriculture are also to be managed with a holistic work plan.

The following salient points have come up, which should be given priority during the course of planning to manage the challenges:

a. Transformation of forest lands to agriculture land should be highly restricted as forests are now not available for conversion.

b. The selection of crop varieties should be very specific to the need of the locales and commensurate with the agro-climatic factors, eco-floristic zone and forest types.

c. Land use planning should be done very judiciously so that the vulnerable natural ecosystem is definitely protected.

d. Terrace cultivation should be done with the help of proper agro-forestry mixing tree crop so that erosion can be checked.

e. Land transformation from intensive mode of agriculture to organic and sustainable mode is urgently required. Although HP, Uttarakhand and Sikkim have already made the plan the other two states are still in dilemma. They are thinking about the HYVs and intensive agriculture to maximize production in a non eco-friendly way.

f. The Hill states of North India must broaden their mind to accept organic farming to conserve the vulnerable ecosystem of the region. However, proper time should be given for the land transformation and the top-bottom approach should be followed. Maintaining strictness in organic farming cannot be overlooked.

g. Optimization and sustainability should be the goal rather than maximization as that would be eco-friendly.

h. More intensive hill specific agri-research plan is need of the day in which two essential components should be (i) region specific variety/provenances/ ecotypes/chemo-types selection and (ii) providing proper market access to the community of the remote and less accessed area.

i. Crop diversification is highly welcome but Government department should be very careful that a good and stable market facility should be made available for the Himalayan farmers otherwise everything will go in vain.

j. Traditional and folk varieties of the crops should be brought into cultivation and application of organic fertilizers management and IPM principles must be followed.

k. Last but not the least Government should prepare hill- or Himalayan specific agriculture policy as the ecosystem is vulnerable and its sustainability is highly important for holistic sustainability of the country.

Conclusion

In a country like India which is dominated by small farmers fighting poverty, agricultural transformation by constantly inserting new agro-technology and its adoption are real difficulties. It is a great achievement that India despite being a developing country, carried on rolling on four wheels of development i.e., technological, human, natural resource and economic development. Farmers need requisite capital to invest in adoption and implementation of modernized agriculture and urgent attention is needed with a holistic and integrated approach. Agricultural scenario in Indian plains during the post-independence era mainly depicted the up-scaling the area under farming and augmentation of production. Establishment of diversified institutions and related infrastructures to accelerate the allied activities was also intensified. Supply of heavy inputs to the farmers, sensitization, creating motivation, skill upgradation, etc., were given priority to reap the maximized benefit from the field. As a result of the country level concerted and integrated effort supported by global intensification in the agri-research aiming at quality and quantity improvement through natural as well as artificial processes using high yielding varieties transformation (traditional to modernized technology intensive) of country-agriculture scenario of a very great magnitude happened which was termed as 'Green Revolution' in the history of Indian agriculture. The input intensive agriculture systematically and scientifically empowered the agriculture sector including the farmers, markets, industries and education in the country level and made the country a real agri-super power in the world. But at the same time due to heavy input of chemical and synthetic fertilizers and pesticides, soil health has been heavily affected and sustainability could not be attained. Therefore, this is the high time to lay emphasis on the sustainability issues of National agriculture.

In the Himalayan region, a hill specific agriculture policy is urgently required keeping in view that the ecosystem is vulnerable and anthropogenic pressure has been increasing. Parallel stress is required to be given on socio-economy for which ultimately the plan would be implemented. The traditional as well as unconventional crops also require consideration besides organic farming. The terrace agriculture requires attention while Jhum or shifting cultivation must be restricted. Region specific agro forestry and farm-forestry models can be good suggestion. Establishment of linkage between high hill farmers with rural agro industry and cottage industry which ultimately can show the silver line of the rural development of Indian Himalayan entire geography and demography as a whole.

It is to be emphasized that the whole country should get the benefits from the National policies. More importance should be given to crop insurance, industrial tie ups, etc., so that farmers would not be the losers because of unpredictable market demands and fluctuating price regimes. Attaining sustainability in agriculture production is a stiff challenge facing a developing country like India mainly as India cannot afford less production. But hopefully, proper region specific plans supported by a strong fiscal policy, administrative efficacy, proper stress on research & development sector considering farmers' socio-economic condition, would surely help the country to find out a sustainable solution, which would be reflected in the country's food and price policies in near future.

(Dr. Basistha Chatterjee and Visvarup Chakravarti are Faculty Members at Rural Management and Business Economics in IBRAD School Management and Sustainable Development, West Bengal, India. They can be reached at basisthachatterjee@yahoo.com and visvarupchakravarti@yahoo.co.in respectively.)

References

Agriculture for Development, "The Agenda for Latin America & Caribbean", World Development Report, 2008.

Bardhan,P.K., "Green Revolution and Agricultural Laboures", *Economc and Political Weekly*, Special Number, 1970.

Barun.J.V., Gulati.A and Fan.S., "Agricultural and Economic Development Stratagies and the Transformation of China and India", IFPRI (2004-2005), Annual Report Essay IFPRI (2004+2005), 2005.

Beckford, G. L., "Transforming Traditional Agriculture", *Comment Journal of Farm Economics*, November 1966.

Bertelsman. S., Gutursloh, "BTI – 2008 – India Country Report", 2007.

Bhalla G.S., "Changing Agrarian Structure in India – A Study of Green Revolution in Haryana", Meerut, Meenakshi Prakashan, 1974.

Bichalo and Hoefle, "Divergent Trends in Brazilian Rural Transformation: Capitalized Agriculture in the Agreste & Sertao of the Northeast", *Bulletin of Latin American Research*, Vol.9, No.1, 1990, pp.49-77.

Boserup, E., *The Conditions of Agricultural Growth*, Aldine Publishing Company, Chicago, 1965.

Breman.J., *Patrouge and Exploitation – Changing Agrarian Relations in South Gujarat, India*, Berkley, University of California Press, 1994.

Carstensen, V. "The land of Plenty" in Issues & Themes, American Association for State & Local History (through U. S. I.S.) New Delhi, 1975.

CMIE, "Basic Statistics Relating to Indian Economy", Vol.I, August 1994.

Deepak.L., "Agricultural Growth, Real Wages and the Rural Poor in India", *EPW*, June 1970.

Dunn, J.M., "Review of Transforming Traditional Agriculture", *Journal of Agriculture Economics*, June,1965.

Gill.S.S., Lindberg. S, Thandi. S. , Babu. K.S., "Panel 32: Post-Green Revolution Agrarian Transformation in South Asia: Ecology and Peasand life Under Globalization", Panel Organization, 2007.

Goodwin, H. G., *Economics of agriculture,* Reston, Reston Publishing Co, 1977.

Google World Atlas, 2007.

Government of India, *Report of the Committee on Unemployment*, May, 1973.

Halcrow, H.G. *Economics of Agriculture,* Mc. Graw Hill International Book Agency.

Hazell, "Transformation in agriculture & their implications for rural development", *Centre for Environmental Policy,* Vol.4 (1), 2007. pp 47-65.

Koushal. N., "Poverty Ratio in Initial Years of Post-reform Era", *Economic Times,* January 11, 1995.

Kuznet, S, *Economic Growth & Structure,* Oxford & IBH Publishing Co., New Delhi. 1965.

Meheran, G.L.(1951), "Comparative Costs of Agriculture Price Support in 1949 Proceedings", *American Economic Review,* May,1951.

Mellor, J.W. and Stevens, R. D. "The Average & Marginal Product of Farm; Labour in Under developed Economics", *Journal of Farm Economics,* August 1956.

Mellor, J.W., *Towards A Theory of Agricultural Development, Agricultural Development & Economic Growth,* London, Cornell University Press 1953.

Metcaff, D, *Economics of Agriculture,* Harmondworth Penguin Books.

Nicholls, W. H., *The place of Agriculture in Economics Development, Agriculture in Economic Development,* Vora & Co. Publishers Pvt. Ltd., Bombay, 1964.

Paul. S., "Green Revolution and Income Distribution among Farm Families in Haryana", *EPW,* December 29-30, 1989.

Planning Commission, *Approach to the Ninth Five Year Plan, 1997-2002,* February 1999.

Rana. T.S.; Dutta. B and Rao., R.R. Flora of Tons valley.. Bisen Singh Mahendra·Pal Singh, Dehradun. 2003.

Rao, C.H.H., *Technical Change and Distribution of Gains in Indian Agriculture,* Delhi, MacMillan Co. of India Ltd, 1976.

Rao.C.H.H., "Technical Change in Indian Agricultural: Emerging trends and perspective", *Indian Journal of Agricultural Economics,* October-December 1989.

Reddy.V. and V.A and Bhaskar. G., *Rural Transformation in India: The Impact of Globalization,* New Delhi, New Century, 2005, XVIII, p.397.

Rosentein- Rodon, P. N. "Disguised Unemployment & Underemployment in Agriculture", *Monthly Bulletin of Agriculture & Statistics,* July- August, 1957.

SAARC Conference on Science-Based Agricultural Transformation towards Alleviation of Hunger and Poverty in SAARC Countries, SAARC Conference Secretariat, Press Release, New Delhi, March 6, 2008.

Sawant .S.D., "Performance of Indian Agriculture with special reference to regional Variation", *IJAE*, July-September 1997.

Schultz, T. W., *Transforming Traditional Agriculture*, New Haven, Yale University Press, 1964.

Schultz, T.W., *The Economic Organization of Agriculture*, Bombay, TMM 1953.

Schultz, T.W. *Agriculture in Unstable Economy*, New York, Mc.Graw Hill, 1945.

Schultz, T.W., *Production & Welfare in Agriculture*, New York, The Macmillan Co. 1949.

Schultz, T.W., *Transforming Traditional Agriculture*, New Haven, Yale University Press, 1964.

Sen.A.K. *Employment Technology and Development*, Oxford, Clarendon press, 1975.

Vaidyanathan. A. "Employment Situation: Some Emerging Perspectives", *EPW*, December 10, 1994.

Visaria.P, "Rural Non-Farm Employment in India: Trends and issues for Research", *IJAE*, July-September 1995.

Vlasin, New Economic Roles for the Rural Environment, Some key issues & challenges posed by non-agricultural Demands for rural development, proceeding paper of winter meeting of the American Agricultural Economic Association with Allied Social Science Associations, Detroit, December 27-29, 1970.

Witt, E, "Role of Agriculture in Economic Development, A Review", *Journal of Farm Economics*, February 1965.

World Bank, "Economic Development in India: Achievement and Challenges", Washington,D.C.USA.

Contribution of ICT in Agricultural Transformation

– Subir Ghosh

With the goal of creating a market oriented agrarian economy agricultural transformation plays a crucial role. The latter is defined as the process of specialized production process from a diversified subsistence one. While ICT integrates the technology and the distribution process to communicate the required information to the right audience to make the latter more participative in nature. Hence to make agricultural transformation a successful one, disseminating right information at the right time unearths the role of Information and Communication Technology (ICT). ICT have addressed problems like cost issues in one to one information dissemination and hurdles in reaching the target audience while implementing agricultural transformation.

In the context of Indian agriculture, application of ICT is manifested through different models like Kisan Call Centers (KCCs), The Gyandoot project, Bhoomi project, and AGMARKNET.

The KCCs was launched on January 21, 2004 by the Department of Agriculture and Cooperation involving technologies like:

- Desktop Computer system with Internet connections.
- Telephones with headphones as well providing the teleconferencing facility.

Through KCCs, extension services were rendered to the farming community with their queries being addressed in their local dialects with virtually zero costs on the part of the farmers. These centers were important information corridors for the Indian farming community.

The Gyandoot Project was an initiative in the Dhar district of Madhya Pradesh covering five lakh people of 311 gram Panchayats, 600 villages and 26 'soochnalayas'. The latter are information centers at the village level in collaboration with the Government of India. The process facilitated the agricultural transformation process by providing information on agricultural produce, availability of landrecords, online registration of applications and auction sites. A concept of village auction site was introduced in 2002 allowing the farmers and the producers to sell of their land, agricultural machinery and equipment, and other durable commodities. Thus these initiatives of ICT in a way also encouraged other allied activities associated with agricultural transformation. These services were provided in exchange for a minimum user fee.

Next is the Bhoomi Project of Karnataka which focused on computerizing the land records throughout the state. This project mainly helped the farmers on land related aspects procuring the Rights, Tenancy and Cultivation certificates. These help the farmers in securing loans from banks, and on line connectivity to courts for usage of landrecords to reconcile civil disputes on land ownership and cultivation.

AGMARKNET (Agricultural Marketing Information Network) is the most important contribution of ICT in the field of agricultural transformation. The marketing aspect of the agricultural transformation is looked upon through this network. It is basically a market information system increasing the efficiency of marketing activities, global information network, stock status, information on market functionaries and facilitating the producers, traders, and consumers.

(Subir Ghosh, Senior Faculty Associate, Icfai Research Centre, Kolkata. He can be reached at subirg@iupindia.org).

12

Contract Farming and Horticulture
A Perspective on Agricultural Transformation in India

Rahul Gupta

Though India is a leading producer of several food items, a majority of Indian farmers, especially the small and marginal ones, continue to be economically backward. The horticulture sector, where small holding farms are common, is ideally suited for contract farming which would be mutually beneficial to both the farmers and agribusiness.

Indian Agricultural Sector

Agriculture is one of the most important sectors of the Indian economy, accounting for about 25% of the national GDP (industry 26% and service sector 49%). The post-Green Revolution has seen India emerge as a leading food producer in the world. The focus of the Green Revolution was primarily on food-grains. With a production level of over 210 million tonnes in 2002, India has emerged as the largest rice producer and the second largest wheat producer in the world. Considerable strides have also been made in other areas of food production like the dairy industry, poultry and livestock and horticulture. While India ranks

---

Source: Icfai Business School Case Development Centre. *This article earlier appeared with the title "Contract Farming: Ideally Suited for the Horticulture Sector" in Effective Executive, June 2005, published by the Icfai University Press.*

first in milk production, it is the fourth largest egg producer and ranks eighth in broiler production. The wide variety of agro-climatic conditions have helped India produce a large variety of horticultural products and India ranks second to Brazil and China in fruits and vegetables production respectively.

In spite of the impressive production figures, a vast multitude of Indian farmers, mainly in the small and marginal category, continue to be economically backward. The small and marginal farmers are increasingly finding it difficult to make their farming venture commercially viable. They also find it difficult to establish links with the marketing system. The land ceiling legislation in post independence India restricted farm size to family owned small farms. Over the years, the number of farms have further increased, while the average size of the farm holding is gradually shrinking. In the mid-nineties, there were 115 million (97.7 million in 1985-86) farm households with an average farm size of 1.5 hectares or less. Table 1 shows the predominance of small and marginal farmers in the landholding distribution in India.

Table 1: Distribution of Landholdings

Landholding	Criteria	Average Size in Hectare	% of Total Number of Holdings	% of Total Area
Marginal	< 1 hectare	0.40	59.0	14.9
Small	1 to 2 hectares	1.44	19.0	17.3
Semi-medium	2 to 4 hectares	2.76	13.2	23.2
Medium	4-10 hectares	5.90	7.2	27.2
Large	> 10 hectares	17.33	1.60	17.4

Source: Ministry of Agriculture (1996).

The number of marginal farmers is on the increase. Farmers with less than 1 hectare holding (operated area: 0.4 hectare), increased from 35.7 million in 1970-71 to 62.1 million in 1990-91. Also, in the small farmer category (<2 hectare), the increase was from 49 million to 82 million during the same time. This vulnerable section accounts for 78% of farmers, up from 69.7% in the mid-seventies. Formulation and execution of proper policies directed towards protecting the interest of the marginal and small farmers should get top priority.

Salvation for Small Farmers

Small farms are becoming unviable due to the inadequate farm inputs and farming techniques available to the farmer. The desired linkages in terms of extension advice, credit, farm mechanization, seeds, fertilizer and agro chemicals on one hand and the assured and remunerative market for the farm produce on the other, can be provided by a well designed contract farming scheme.

In contract farming, the agribusiness or the sponsor gets into an arrangement with a group of farmers to grow a variety of crops on an agreed area with the assurance of buying back the produce at a predetermined or negotiated price. The sponsor usually supplies the farm inputs, extends credit, provides extension services and upgrades farmers' skills on advanced farm practices. The farmer provides the land and labor. Thereby, while small farmers can make farming commercially viable, contractors or agribusiness can be assured of a consistent source of supply, in terms of quality and quantity.

Contract farming binds the farmer and the sponsor in a business deal with a win-win situation for both the parties, where the projects are not motivated mainly by the political or social considerations, like in the Green Revolution, but is tied up with economic and technical realities. The contract is drawn up between unequal parties—the agribusiness, entrepreneurs and government agencies on one side and the economically weaker small farmers on the other. The design should aim at reducing the risk of either party as compared to buying and selling of agricultural produce in the open market.

Contract Farming Model: The Advantages and the Problems Faced

Advantages

In order to be sustainable, contract farming should offer benefits to the farmers and sponsors. Proper design should tackle the problems that exist along with these advantages. The farmer enjoys benefits like access to inputs and production services, credit, appropriate technology, acquiring advanced farming skills, predetermined remunerative price structure and access to markets. Left on their own, the small farmers may not find access to credit to finance the inputs. The sponsor directly, or through arrangements made with commercial banks, provides the required credit and the contract serves as the collateral. Small farmers are reluctant to adopt new technology, which is often a necessary requirement for achieving quantity and meeting the quality standards of the market. The farmers are more prone to adopt the risk of new technology when they have the support of the extension services provided by the sponsor. The skills transferred to the farmers improve their overall farm management and improves utilization of farm resources and inputs. In the open market, the small farmers have to negotiate with prospective buyers or go by the market prices. Considering the low economic strength of small farmers, they often lose out on these deals. Contract farming can protect small farmers from such loss. Finally, farmers are assured of a market for their produce as the contract specifies that the sponsor will take the agreed output. Often the product is lifted at the farm doorstep.

The sponsors too can derive several advantages from the contract farming model. The major advantage is the reliability of the product in terms of quantity and quality. Sponsors who have invested considerable resources in setting up processing plants require assured raw material to keep these plants running. Neither is purchase from open market reliable, nor is it always possible to own plantations. Contract farming provides sponsors the required amount of products and the right quality. For agribusiness involved in processing, uniform product collection of a particular variety is required. Through contract farming, uniformity of seeds distributed to individual farmers and standardization of the farming methods can guarantee this product consistency. Finally, farm input manufacturers can capitalize on the contract farming model for promotion of their product. An agrochemical manufacturer may join hands with a food processor or an exporter to promote a contract farming model involving small farmers.

Problems

The farmers who get attracted to contract farming due to the easy access to credit provided by the agribusiness or through third party tie ups are likely to face indebtedness in case of a crop failure. Crop failure may be due to the faulty extension services offered by the sponsor. Even with production being on target, changed market conditions or the sponsors' failure to pick up the produce may put the farmer in a similar spot. For tree crops, like tea or rubber, where the farmers are unable to change to alternative crops easily, the sponsor may enjoy a monopolistic situation and exploit the farmer. The adoption of advanced farming technology and innovations introduced by the sponsors must be compatible with the competencies and the social norms of the farming community.

The sponsor can also face some typical problems. A major problem can be the sale of produce by the farmer outside the contract, at a higher price, in case of a shortage situation. Diversion of inputs supplied by the sponsor for uses other than what has been intended under the contract is another problem faced by the sponsor. Farmer discontent arising out of delayed payments, deviation from contracted prices, inadequate extension services, etc., may erode the confidence of the farmers and they may decide not to participate in such a venture in the future.

Despite these problems associated with contract farming, the advantages are strong enough to adopt this model for improving the efficiency and productivity of small farms in India. Agribusiness, both under the corporate banner as well as farmers' cooperatives and government agencies can partner with small farmers in order to effectively consolidate the fragmented land holdings and have cost effective agricultural practices through the contract farming model. Fortunately, this model has been tried out in some isolated instances in Indian agriculture with reasonable success.

Contract Farming in India

Contract farming is not new to India. As early as 1920s, ITC started cultivation of a variety of tobacco in Andhra Pradesh that was similar to contract farming. Again, in the 1960s seed companies essentially produced seeds on several contracted individual farms, as the seed producers do not own land. Co-operatives for milk (in Gujarat) and the sugar cane (in Maharashtra) that came up during that time, followed the framework of contract farming, where it bought the members' products at contracted prices. But the spread of contract farming has been slow till it gathered momentum recently with rise in value-added agricultural export and liberalization.

Agribusiness major Pepsi Food Ltd., met the requirement for its Hoshiarpur (Punjab) based tomato processing plant, to produce aseptically packed purees and paste, through a successful contract farming model. The quantity required for the processing plant (40,000 tons) was far in excess of the local production (28,000 tons). But till that time the farmers had to throw away their produce for lack of buyers. Through use of standardized seeds, proper extension services and monitoring, the crop yield was raised to the targeted level and a productivity of 52 tons/hectare (up from 14-16 tons/hectare) was achieved. In trial chilli contract farming also they increased yield significantly from 6 tons/hectare to 20 tons/ hectare. The farmers' income increased, while the sponsor was able to keep his processing plant busy. Pepsi Foods Ltd. did not go about this project in isolation. They entered into a partnership with Punjab Agricultural University and Punjab Agro Industries Corporation. By aligning itself with local bodies, Pepsi, despite being an outsider and a multinational could easily win the confidence of the farmers. Among other vegetables, they successfully included potato, to meet the demand for raw material for their potato chips business. Their commercial success in establishing the backward integration through the contract farming model in tomato cultivation prompted them to extend the practice to other crops including food grains, spices and oilseeds. Pepsi followed the ideal path of setting up processing plants near the farms to add value to the farm produce.

Some of the key elements in Pepsi's success in tomato contract farming were:

- Effective R&D through a dedicated team.
- Effective extension services where trained experts transferred technology to farmers.
- Involvement of agricultural universities and public sector units like Punjab Agro Industrial Corporations, which enhanced the acceptability of Pepsi among the farmers.
- Timely supply of farm inputs, including implements.
- Availability of credit.
- Timely payments to farmers, and
- Prompt procurement of produce and implementation of proper logistic system.

In the area of foodgrains, export potential of the famous basmati rice has drawn several exporters towards practice of contract farming in order to ensure quality and quantity required by the overseas market. Pepsi, after conducting field trials, in 2002-03, initiated the project to grow basmati rice over about 800 acres of farmland in Punjab and Western Uttar Pradesh in the contract farming model. It had plans of expansion to 4,000 acres by 2004, to make its processing plant run to full capacity.

Not only the exporters, we also have instances of manufacturers of agricultural inputs tie up with the exporters, government agencies and commercial banks to promote contract farming jointly. One such project teams up Punjab government, with agrochemical manufacturer Rallis India Ltd., ICICI Bank and LT Overseas Ltd., for practice of contract farming to grow basmati rice over 30,000 acres. Another manufacturer of agricultural machinery, Escorts Ltd., has entered into an agreement with Punjab Agro Industries Corporation, to develop the contract farming initiatives spread over about 50,000 acres for basmati, wheat and oilseeds, during 2003-04. Yet another exporter of basmati rice, Satnam Overseas Ltd., planned to procure 30% of about 100,000 tons of basmati rice that it sells in the domestic and the export market through contract farming arrangements with farmers' associations in Punjab and Haryana. Contract farming can be an effective means to achieve crop diversification for which Punjab has ambitious plans to cover about 25% of its total 10.5 million acres of cultivable land. In the food grains area, Hindustan Lever Ltd., a leading player in processed food, has successfully used contract farming to grow wheat in Madhya Pradesh, for its raw material requirement in its branded flour production.

Not only food crops like vegetable, grains and oilseed, but also cash crops like cotton has seen some form of contract farming in Appachi Cotton Company's venture to organize farmers in Tamil Nadu to grow cotton against a market assurance to buy the product. Also, Uger Sugar Works Pvt. Ltd., in Karnataka, fed its malt production plant, located close to the barley growing area, by sourcing a part of their raw material requirement from a contract farming venture.

These are some instances of successful contract farming in India. The account is not exhaustive, but it provides an insight into gainful adoption of the concept of contract farming in Indian agriculture.

Scope of Contract Farming Application in India with Special Reference to Horticulture

We have seen that contract farming has had successful applications in consolidating many small farmers and getting the raw material required for the processing plants. In the Indian context, horticulture provides an ideal backdrop for practice of contract farming for selected crops. Horticulture accounts for about 28% of GDP in agriculture while the area under cultivation in horticulture is only 8.5% of the total cultivated area. Horticulture production also accounts for 52% of export share in agriculture. Growth rate in export has been 12% in quantity and 34% in value between the period 1991-92 and 2002-03. A wide range of agro-climatic conditions prevailing in India has made it possible to grow a variety of horticultural products. The Green Revolution primarily focused on food grain and horticulture cultivation continued to remain fragmented with majority of the farmers being in the marginal to small category. In spite of this, horticulture production has seen healthy increases. From 96.56 million tons in 1991-92, the production has gone up to 145.78 million tonnes in 2001-02. The annual growth rate for horticulture has been 5.7% as against a growth of 2.9% for food grains. The National Horticultural Mission of the government has set a target of 300 million tons by 2011-12. The productivity of vegetables has increased from 10.5 tons per hectare in 1991-92 to 14.4 tons per hectare in 2001-02. For fruits, these figures are 10 and 10.7. Though there has been a productivity increase of about 17.3% during the period, overall horticultural productivity is about 0.62% that of China and Brazil. Lack of storage and processing, results in a loss of around 30% of the output. As of 2002-03 less than 2% of the fruits and vegetable produced is processed, utilizing 46.04% of the installed processing capacity. The mission also proposes to increase the area under cultivation which may require shifting land under cultivation from the traditional crops like rice and wheat. The production and area of key vegetables and fruits are given in Table 2.

Notwithstanding the comparatively low acreage under horticulture, value wise contribution of horticulture is considerable. With value addition through processing, the economics of horticulture cultivation can be further improved. Processed horticulture products can find easy export market. Agribusiness can play a crucial role here by setting up processing plants close to the production zone and organizing the farmers in the contract farming model.

Table 2: All India Area and Production of Major Horticultural Crop (Area: Million Hectares, Production: Million Tons)

Crop	1999-2000		2000-01		2001-02		2002-03	
	Area	Prdn	Area	Prdn	Area	Prdn	Area	Prdn
Fruits	3.80	45.50	3.89	43.14	4.00	43.00	4.18	47.68
Apple	0.23	1.04	0.24	1.23	0.24	1.16	0.25	1.47
Banana	0.49	16.81	0.47	14.14	0.47	14.21	0.68	16.82
Mango	1.48	10.50	1.52	10.06	1.58	10.02	1.60	10.78
Citrus	0.52	4.65	0.50	4.40	0.62	4.80	0.60	4.72
Vegetable	5.99	90.83	6.25	94.00	6.20	88.62	7.59	97.50
Brinjal	0.50	8.12	0.47	7.70	0.50	8.35	0.50	7.83
Cabbage	0.26	5.90	0.25	5.51	0.26	5.68	0.28	5.80
Cauliflower	0.25	4.72	0.26	4.69	0.27	4.89	0.28	4.80
Okra	0.35	3.42	0.35	3.34	0.35	3.32	0.37	3.53
Onion	0.49	4.90	0.45	4.72	0.50	5.25	0.53	5.45
Pea	0.27	2.71	0.32	3.01	0.30	2.04	0.35	3.20
Potato	1.34	25.00	1.21	22.44	1.25	24.45	1.37	25.00
Tomato	0.46	7.43	0.46	7.24	0.46	7.46	0.54	7.60
Flowers	0.09	0.52	0.10	0.56	0.11	0.54	0.15	0.70
Cashew	0.69	0.52	0.72	0.45	0.74	0.46	0.73	0.47
Coconut*	1.77	12.23	1.82	12.68	1.89	12.82	–	–

*Production in thousand million nuts.

Source: Indian Horticultural Database 2002-03, Ministry of Agriculture.

Horticulture products coming out of organic farms can also find a good export market in the estimated $37 bn worth market for organic food in 2005 ($17 bn in 2000). India's share in this market is about 0.001%. In India a large number of small and marginal farmers who do not have access to chemical inputs, effectively follow the principles of organic farming. Ironically, this production does not pass off as products of organic farming due to the absence of certification. The government has set up the National Institute of Organic Farming in October 2003, with an allocation of Rs.100 cr for the promotion of organic farming in the 10th five year plan. With this potential for organic products, agribusiness and exporters can organize the small horticulture growers, with the help of the National Institute of Organic Farming, for large scale organic farming output, in the contract farming model.

The multitude of fragmented farms producing horticulture products need to be organized in order to provide each individual farm the strength to sustain and give the economy of scale for growth and profitability through consolidation of several small farms. The participation of industry, mainly firms involved in agribusiness, will provide strength to these small farmers, organized in self-help groups, in terms of access to inputs and also help in setting up direct linkage with the marketing channels for their produce. Development of supermarket chains will also help in such chains' backward integration to tie up farmer groups to market their produce directly through their stores under contract farming agreement. The government support in terms of policy formulation to promote contract farming has been taken up by some states.

Government Role in Promoting Contract Farming

The government can extend fiscal support and incentives to agribusinesses that are coming forward to organize the small farmers through contract farming ventures. These incentives can be on a time frame to help the agribusiness and farmers during the initial years when the model gains acceptability among the farmers. These benefits can be in the form of tax exemptions/ rebate from income generated from such ventures. Also, import duty on processing plants and machinery can be waived. A legislative procedure can be drawn up to ensure enforcement of contract terms and conditions, both on the part of farmers as well as the sponsor and for the quick settlement of disputes between the farmers and the sponsors. Appropriate crop insurance policies can be promulgated to protect the parties involved in the contract. The government should provide all support to the private sector in development of farm technology development centers for spread of modern farming methods. Government agencies should recognize the system of contract farming by including such schemes while developing cropping plan specific to a region and propagate it through agricultural university extension services.

The government should garner its resources in setting up the necessary infrastructure for cold chains, refrigerated trucks and warehouses. Government agencies that monitor agriculture and agronomy should also ensure that intensive farming resorted to by agribusiness does not drain the soil of its nutrients and ground water.

Conclusion

A small farmer, who often does not get remunerative returns from his farming operations, with the middleman eating into the farmers share, is more likely to get his due share through a contract farming arrangement. On one hand, he gets access to world class farming technology, in terms of seeds, agronomic practices, farm implements, etc., while on the other, he is assured of a predetermined price and a market outlet at his farm gate. With upfront assured prices, the focus shifts to improvements in productivity, arising out of the constant monitoring of the sponsors agronomy specialists. Another major benefit in such contract farming arrangements is that the tie-up is on a long-term basis, as opposed to the short-term nature of an open market price support system. This effective partnership between the private and corporate sector and small holding agriculture is increasingly finding application in the form of contract farming to overcome the problems posed by the small holding size of Indian agriculture.

(Rahul Gupta, Faculty Member, The Icfai Business School,, Kolkata. He can be reached at rg1609@rediffmail.com).

Impact of Contract Farming in Agricultural Transformation

– *Subir Ghosh*

Contract farming is an agreement signed between the farmer and the land owner for a specific period of time to produce agricultural crops. Both the parties are benefited under contract farming because both the parties think about quality product, risk reduction and political authentication. Under contract farming, the land owner generally provides good quality of mother seeds, farming equipments, new technology and also helps the farmer to dispose the products at higher price. On the other hand, farmer concentrates on how to produce qualitative crops. In return to that the land owner/contractor shares the profit with the farmer as per agreement and generally this return becomes higher than what the farmer earns separately.

Agricultural transformation refers to the process of specializing and making the agricultural produce towards-market oriented. Now the question arise which method or technique will be adopted to make agricultural sector more productive and at the same time qualitative production because qualitative production will only can ensure guarantee to face international competition and help to sell the agricultural produce at high prices. It also needs wide market opportunities to sell agricultural produce. To get maximum yield from agricultural sector, the involvement of agri-business houses are also required as they can support to the marginal farmers through the provision of qualitative mother seeds, fertilizers, etc. and also help the farmers to dispose agricultural produce at best market prices. Therefore, contract farming is one of the best options for achieving agricultural transformation.

According to the Food and Agriculture Organization (FAO), "HLL's contract farming resulted in 64% higher yields by the farmers, which was shared mutually. Pepsi Co's farming contract saved 20%-30% yields, which was split in half with the farmers: that actually doubled their incomes. Further the end consumer was also benefited due to a 10% saving on supply chain efficiencies and economies of scale."

Therefore, the role of contract farming cannot be ignored to achieve successful agricultural transformation of a nation.

(Subir Ghosh, Senior Faculty Associate, Icfai Research Centre, Kolkata. He can be reached at subirg@iupindia.org).

References

Ghosh Subir (2008), "Alternative Farming: An Introduction", published in *Alternative forms of farming in India*, The Icfai University Press.

http://www.isec.ac.in/CONTRACT%20FARMING%20FINAL%20REPORT.pdf

http://www.fao.org/ag/ags/AGSM/contract/cfmain.pdf

INDEX

D

E

F

G

H

I